Corporate Technological Behaviour
Co-operation and Networks

Efficient technological strategy is an increasingly important element in industrial profitability. An understanding of networks – the formal and informal web of contacts between suppliers, producers and customers – is vital to the application of such strategy. In this book Håkan Håkansson brings together theory and practice to provide the first comprehensive and detailed study of technological development in companies, and the associated interactions with other companies and organizations.

Beginning with an account of the theoretical frame of reference, Håkansson goes on to examine corporate technological behaviour by focusing on the technical, purchasing, selling, personnel, and financial characteristics of over 120 companies. This detailed case-study research gives much valuable information about the nature of the network interaction and the form that it takes, enabling readers to analyse its effects both at the corporate level and in society more generally. The author also discusses the broader implications of these case studies, especially for business managers and decision-makers concerned with industrial policy, and indicates how managers and others can improve technological research by exploiting networks efficiently.

The Author: Håkan Håkansson teaches in the Department of Business Studies at the University of Uppsala, Sweden.

Corporate Technological Behaviour
Co-operation and Networks

Håkan Håkansson

Routledge
London and New York

First published 1989
by Routledge, 11 New Fetter Lane,
London EC4P 4EE
29 West 35th Street, New York, NY 10001

© 1989 Håkan Håkansson
Typeset by Witwell Ltd, Southport
Printed in Great Britain by
Billing & Sons Ltd, Worcester

All rights reserved. No part of this book may be reprinted
or reproduced or utilized in any form or by any electronic,
mechanical, or other means, now known or hereafter
invented, including photocopying and recording, or in any
information storage or retrieval system, without permission
in writing from the publishers.

British Library Cataloguing in Publication Data

Håkansson, Håkan, *1947–*
   Corporate technological behaviour
   1. Companies. Computer systems. Networks
   I. Title
   658'.05
   ISBN 0-415-00020-3   Routledge

Library of Congress Cataloging-in-Publication Data

Håkansson, Håkan, 1947–
   Corporate technological behaviour
   Bibliography: p.
   Includes index.
   1. Technological innovations. 2. Research, Industrial.
3. Cooperation. I. Title.
T173.8.H35   1989   607'.2   88-32537
ISBN 0-415-00020-3

Contents

List of Figures	vii
List of Tables	viii
Preface	xi
1 Technological development and the company	1
2 The company in a network	15
3 Technological development in industrial networks	29
4 Empirical study	44
5 Corporate network behaviour	51
6 Resource structure and corporate network behaviour	65
7 Corporate activities and network behaviour	99
8 Individual development relationships – characteristics of the partners	108
9 Technological co-operation and corporate relationships – summary and theoretical implications	120
10 Network analysis for corporate management	131
11 Network analysis for industrial policy	152
Appendix: Questionnaire used in company interviews	173
Bibliography	201
Index	207

Figures

2.1	Network model	17
2.2	A network of six actors	19
4.1	Variables studied	46
5.1	Expected short-run effects on profit of different relative shares of external collaboration	61
5.2	Long-run effect on profit of different relative shares of external collaboration	62
6.1	The theoretical marketing network	66
6.2	Number of co-operation relationships with the ten biggest customers	67
6.3	Technological adaptation to customers	69
6.4	Volume purchased in relation to turnover	76
6.5	Number of co-operation relationships with suppliers	78
6.6	Degree of concentration on the customer and the supplier sides	79
6.7	The company and the capital network	83
6.8	Personal networks	85
6.9	Number of horizontal technological development relationships	93
6.10	Corporate collaboration partners in technological development	98
9.1	Triangles: balance and imbalance	129
10.1	Checklist for developing a corporate identity	134
10.2	Theoretical structure of the hydraulics network	137
10.3	Resource base of a steel company over a 75-year period	143
10.4	Matrix for monitoring technological development in corporate networks	144
10.5	Knowledge transmission channels and their effects	145
11.1	Type of units included in the biotechnical network and the most common relations between them	154
11.2	The genetic engineering network	156
11.3	The mechanical engineering network in Uppsala County in a technological development perspective	167

Tables

2.1	Resource control	18
4.1	Companies studied	47
5.1	Relative share of development activities conducted in collaboration with external units	54
5.2	Characteristics of purchasing and marketing in companies with different co-operation profiles	56
5.3	Profit at different relative shares of external technological development	62
5.4	Volume growth at different relative shares of external technological development	63
6.1	Technological issues on at least 'test level' in relation to other customers	68
6.2	Relationship between number of potential customers and relative share of external technological development	71
6.3	Number of customers in relation to the selling company's external orientation	71
6.4	The effect of market share on the selling company's propensity to co-operate	73
6.5	Relationship between the seller's relative share of external technological development and the number of customers (among the ten biggest) that also buy from the main competitor	74
6.6	New products over a five-year period as a percentage of total purchases: materials, components, and equipment	81
6.7	Ownership and the relative share of external technological development	83
6.8	Relative share of external technological development and number of owners	84
6.9	Relationship between number employed and relative share of external technological development	86

6.10	Relationship between relative share of external technological development and proportion of white-collar workers	87
6.11	Relationship between internal training activity and collaborative relationships	88
6.12	Average length of employment of blue-collar workers	89
6.13	Perceived changes in competence level in four corporate functions	90
6.14	Staff mobility	91
6.15	Shortage of engineers and skilled workers	91
6.16	Capital intensity and relative share of external technological development	95
6.17	Technological content of the product and relative share of external technological development	96
7.1	Total development investment and the relative share of external collaboration	101
7.2	Relative share of external technological development in relation to proportion of customer-led production	103
7.3	Use of distributors	105
7.4	Correlation between use of distributors and technological development in customer relationships	106
7.5	Development relationships with small customers and the use of distributors	106
8.1	Geographical dispersion of different types of partner and of total sales or purchases	110
8.2	Duration of development relationships in relation to type of partner	112
8.3	Form of co-operation	113
8.4	Type of development co-operation	114
8.5	Number of people having direct contact with partner in different development relationships	115
8.6	Frequency of contact (personal meetings)	116
8.7	Results of the collaboration	117
8.8	Expectations regarding the results of established collaborative relationships	118
11.1	Frequency and importance of different channels to the dissemination of information and knowledge in the six general programmes	161
11.2	Relationships with a technological content and location of the partner	168

Preface

A Swedish author, Göran Palm, has claimed that 'Everything reappears in the same self-evident way that everything changes'. This is obviously true when analysing networks as well as when writing. The processes within an industrial network are characterized by both stabilizing and changing forces. Stability and change are in this way interrelated and probably dependent on each other. The same applies to the content of this book in relation to what I and others within 'the relationship and network paradigm' have written before. The stabilizing effect of relationships and networks reappears, but at the same time their development force is emphasized more strongly. Both relationships and networks are, in this context, structuring devices directing resources in specific ways. Industrial networks include large companies and thus these networks are extremely powerful and important to influence and control. They are forming our future.

A book about co-operation and networks is, of course, developed through co-operation and use of a personal network. I have been lucky enough to have been a member of a close research network during the last fifteen years. The core of that network from my point of view is the IM-group in Uppsala. In this specific research project I am deeply grateful to Elisabeth Brandel, Hans Henriksson, Jan Hedman and Alexandra Waluszewski who not only were responsible for the data collection but also were active in designing the project. In the same way Nazeem Seyed-Mohamed and Barbara Henders have taken an active part in the data analysis.

Different versions of the book have been presented in seminars where, among others, Kerstin Sahlin-Andersson, Mats Forsgren, Lars-Erik Gadde, Jens Laage-Hellman, Björn Axelsson, Jan Johanson and Lars Engwall have contributed in such a way that the publication date has had to be postponed in order to adapt to their constructive criticism.

The project, furthermore, has benefited from being a part of a

Preface

research programme named 'Marketing and Competitiveness' headed by Lars-Gunnar Mattsson of the Stockholm School of Economics and financed by STU (The Swedish National Board for Technical Development) and MTC (The Marketing Technology Center).

The book has been translated by Nancy Adler and Inger Håkansson has been responsible for typing and design.

To all of you and to all participating companies and others taking part in this research I can only express my gratitude and my hope that in the future you will find some way to use me in your network as I have used your abilities during this project.

Håkan Håkansson
Uppsala, Sweden

Chapter one

Technological development and the company

On Friday, 13 April 1987, Aerotronics Ltd had to deal with a number of technical issues. At a meeting of the board, for instance, the vexed question of further investment in a major product development scheme was to be discussed. This project had been running for over three years. Apart from Aerotronics itself, it had involved a university department and two suppliers and had already cost a great deal more than originally intended. The results so far achieved promised both well and ill. It now appeared that the potential was considerably greater than had at first been thought, but certain technical problems were also proving more intractable. Debate on the subject had been lively; there was a strong lobby in favour of discontinuing the project, while another thought it should receive more funds. A compromise was devised by the chairman of the board whereby the project was to continue, but on a very much smaller budget. The decision was motivated on the grounds that the project had become something of a cuckoo in the nest, and that halving its budget would reduce it to the same fledgling status as the other projects. Against this it was argued that although the money would permit the project to continue, the whole scheme would be greatly delayed and would lose most of its motivation.

On the same day a salesman and an engineer from Aerotronics visited one of the company's established customers in West Germany. The customer had asked them to come because a number of technical problems had arisen in production. This customer buys advanced material from Aerotronics, material which is then processed and combined with other materials and components to make a fairly unsophisticated final product. The production process in itself is fairly complicated, but can also be regarded as relatively familiar. Over the last few years the customer had exchanged some of the components that were previously included, and had also gradually been replacing one of the other materials. It was when the new material was combined with the Aerotronics product that the trouble arose. The product itself was satisfactory, but the customer's production machinery kept going wrong. The Aerotronics engineer, who thought at first that it was a relatively simple problem, became increasingly puzzled. He made a couple of tests, and the two worried men then returned home. Looking back we now know that these events were to have a positive outcome. In the course of the laboratory tests which were made over the next two weeks, it turned out that they way in which the customer was combining the two materials was producing a particular chemical and electrical effect that was previously unknown, at least to Aerotronics. This effect was causing certain specific problems in the production process, but it also possessed the potential for positive effects on the product. Within a month of the visit to the customer, Aerotronics had succeeded in getting other customers interested in participating in a development project based on the discovered effect.

Apart from these two major events several other happenings of a technical nature occurred on 13 April, which in the slightly longer term may prove to have been important. For example, two of the company's development engineers attended a conference on possible future production methods in the field in which Aerotronics operates. They found the various lectures dull and did not get much out of them, but the conference gave them an opportunity to meet some old friends and even to make a few new acquaintances. What was particularly pleasing, and perhaps also more important, was that they met someone they both happened to know without having realized it before, and that their mutual friend had recently joined one of their oldest customers as a development engineer. All three agreed that the coincidence was worth exploiting.

On the same day various people from Aerotronics accounted together for ten contacts with suppliers. In three cases service personnel from the suppliers visited Aerotronics to examine and adjust various machines. In four cases groups from the suppliers came for an annual review of technical and commercial questions; in two of these last cases, the companies had certain common technical activities which needed investigation. In two more cases members of the Aerotronics staff visited suppliers. One of these was mainly a courtesy visit, although it included a technical review of the supplier's plant and products; the other had a more specific purpose following a number of problems in deliveries, product performance, and the handling of complaints. The tenth and last case was a meeting at a customer company, which Aerotronics had invited one of its suppliers to attend as well. The customer had several ideas (far too wild, in Aerotronics' opinion) about some product changes that would affect both Aerotronics and this particular supplier. To some extent Aerotronics was using the supplier as an external assessor of the ideas being floated by the customer. The supplier was not impressed, and the Aerotronics representatives decided to lie low in the future.

Friday 13 April was not a unique day at Aerotronics; it will certainly not go down in history. Like all manufacturing companies Aerotronics comes up against a variety of technical problems and opportunities every day. Some are purely internal, but many involve interaction with various other parties in the world outside. Some of these events are thought to be important at the time but eventually prove of no consequence. In other cases the reverse is true, an event seems trivial when it happens but in the long run has important implications. Yet others, of course, are regarded as important and prove to be so; others are regarded as less important and again this proves to be the case. But many events that are insignificant in themselves can ultimately involve major changes when taken together.

The company is exposed to a constant stream of events because it is operating in a technologically changing environment. All around it changes of every kind are occurring. The customers and consumers of the company's products are continually changing – in their technology, in their knowledge, or in some other way – and therefore have new and different requirements regarding the products and services they need. A change may mean that customers want more technically advanced solutions, but very often it may mean just the opposite, for instance when customers learn to handle a particular technique and therefore do not need as much technical service as before. Thus the demands on a company can either increase or diminish.

Similarly, changes occur in supplier companies. They may be able to provide products with new or better features which can directly affect the internal production arrangements or the finished products of the buyer company. Again, developments in research institutes, in various regulatory authorities, in the companies manufacturing complementary products, or in other similar units can all affect a company's situation in vital ways. Finally, it is obvious that technological developments in rival companies that make similar or substitute products can have direct implications for a company. Taken together, the many technological developments in the companies and other organizations that make up the contact network of a particular enterprise can affect the situation of this enterprise both by making new demands and by creating opportunities for technological development.

Added to these changes are internal technological developments

Technological development and the company

affecting production processes and products, and these in turn help to determine what units constitute appropriate and/or necessary elements in the contact network, and what the nature of the contacts should be.

The development situation at the corporate level

As our brief introduction has shown, Aerotronics, like all other manufacturing companies, continually encounters a great many different development opportunities. Representatives of the company come up against a variety of problems and openings, to which they have to respond constructively. Let us first try to describe such situations in general (i.e. the problems and opportunities), and then see how they can best be classified and handled.

Several obvious factors emerge as affecting the individual company. We can begin by identifying these, and then try to group them together under a few main headings.

One feature that is immediately evident is the variety of both problems and opportunities. So many different types of situation can arise, and they all have to be dealt with; the problems may concern technical plant, applications, knowledge, people, contacts, and so on.

A similar feature, and equally typical, is that problems and opportunities may occur – or may first be noted – in all sorts of different places: perhaps internally, in a production department or at a drawing-board or during a meeting, or externally on one of the numerous occasions that company representatives meet the representatives of various other organizations, perhaps in discussion or collaboration with a customer, a supplier, or a research institution.

A third typical feature is that many problems and opportunities affect not only the particular company but also some of its opposite numbers outside the company. If a problem is to be solved or an opportunity exploited, changes will be necessary not only in the company but also in one or more of these other organizations. In other words, there is a strong element of mutuality in both problems and solutions.

A fourth feature is that problems and opportunities are often observed at one place, but their solution or exploitation require changes somewhere else. For example, a problem is noted in relations with a customer, but the solution calls for some alteration in relations with a supplier. Production problems or production opportunities may necessitate a change in supplier or customer arrangements, and so on. Thus there are many dependencies between different situations and between the different places where the situations can arise. Added to this, the solution to one problem often conflicts with the solutions to others, thus creating yet another problem that has to be tackled.

A fifth typical feature arises from the dynamic nature of the process in which the company is engaged: all changes occur in relation to other changes, and in relation to other units that are also undergoing change. It is like walking round a room in which the floor leads in one direction and the walls in another, making it extremely difficult to judge the direction of any movement in relation to any other.[1]

A situation characterized by all these features may appear at least superficially to offer such an abundant variety of problems and opportunities that confusion and irresolution must be the natural reaction. And so they would be, in any company that tried to find some total solution to all the problems at once. This sort of approach is almost bound to fail, and such situations are handled instead in a process that can be described in the classical terms introduced by Cyert and March (1963), that is sub-optimization, local rationality, limited search processes, and so on. But – and here we make an important addition – the actors also try systematically to relate the different situations to each other. By exploiting situational interdependencies, various types of advantage can be created.

Anyone working in industry knows that a problem can never be solved exclusively on the local plane; every solution also has implications elsewhere. This makes it possible to influence things in an indirect way, provided those concerned understand something of the relationships involved. In other words there is a great need for mental models and frames of reference to help managements, for example, in their thinking and acting. We have only to consider all the consultants and management books that are bought to help people create suitable frames of reference to realize just how widespread and deeply felt this need must be.

No frame of reference can cover all aspects of events; and since (at least according to the approach adopted here) there are no independent laws behind events but only the laws created by the actors themselves, no-one can claim that his own particular frame of reference is the only right and true one.

A criticism that applies to many of the frames of reference popular today is that they rarely try to capture relationships and dependencies between different situations, different companies, and different types of solution.[2] In the technological field this is a particularly unfortunate weakness, since the technical system itself embraces many obvious interdependencies.

We thus have no model for connecting different situations or identifying the relationships between the various companies and other organizations involved. A reference frame of this kind is certainly needed to enable us to handle two fundamental questions, which will be providing us with the material for our argument in the present book.

The first question is: how are companies to exploit for their own benefit the technological developments occurring in companies or organizations in their environment? All technical changes generate numerous opportunities for changes or adaptations in other organizations. For instance it may become possible for a company to organize production more efficiently, either in the production system itself or even in its link-up with the production systems of suppliers and customers. And obviously the same applies if the company has generated a change in its own product development; such changes should be systematically related (i.e. synchronized or intentionally not synchronized) to external technological developments. It applies also to the extent and focus of the total package on offer to customers, that is the way the product is combined with various types of service. It is just as common in this connection that an existing service is discontinued as it is that a new service is added. A company can be innovative by combining its product with various services, but it could be equally innovative to launch a 'naked' product or to sell a specialized product in a new standardized form. This last type of innovation normally occurs when the customer's technical competence seems so far developed as to render highly qualified back-up service unnecessary or unjustified. This fundamental issue entails a number of other interlocking questions. To which of its co-actors should the company adapt? What technical trends should it follow up? How is it to pick up change signals? Who should be responsible for internal co-ordination?

The other fundamental question concerns the reverse process, that is how the company is to initiate the kind of technological development in other units that will favour its own overall development, for instance as regards the sale of its products and services. This sort of active influence may involve internal technological development or collaboration with one or more external units, but not necessarily so. Nor is active influence the prerogative of large companies with big resources; on the contrary it can be particularly important in small companies with limited resources, since we are not talking only about changes of great general importance but also about various quite specific adjustments. In other words it may be a question of working for the acceptance of whole technological system solutions, but more frequently it is a matter of small adjustments, like combining two products.

Important interlocking questions will then concern who should be influenced and who should be responsible for internal co-ordination, how to set an appropriate course, and so on.

The development problem at the societal level

The complexity of the problems at Aerotronics are as nothing compared with the complexity that prevails at the societal level. In the region where Aerotronics' main operations are located there may be ten or more equally large companies, each one facing similar technological situations, and in the whole country perhaps 100 or even 1,000 such companies. Thus on any single day thousands of situations arise in which technical problems are tackled and various combinations of old and new solutions are tried. Problems, technologies, and solutions are combined on such a scale as to make any all-embracing overview impossible to acquire, which obviously also makes it impossible to plan for such complex variety at either the specific or the more far-sighted level. And yet the process is not altogether automatic or self-operating, nor is it inaccessible to influence. Thus there is every reason and ample opportunity for organizations with some responsibility at the interlocking level to engage in this development, as in fact they always have done.

Originally the development problem was associated primarily with certain countries whose overall level of development was low. Over the last 20 years or so the issue has gradually become increasingly important even in what we like to call the 'developed' countries. At first it was largely a matter of dealing with regional differences. All developed countries have regions that have fallen behind and that have tried in various ways to speed things up. Recently, too, national ambitions of the various developed countries have become more significant. There seems to be a growing fear of lagging behind, or being left out. Up to a point this fear may be part of a generally tougher climate in the countries concerned, but it is certainly also connected with the enormous pressure of technological development particularly in the high-technology fields: this requires such vast resources and has so many facets that the co-ordinated efforts of a great many different types of unit are needed (research institutions, business enterprises, and other organizations) if any real results are to be achieved. Another reason that may be important, particularly in combination with this last factor, is that countries are becoming more closely linked, so that relative differences have more impact than in the past.

One topic that appears in all development theories, whether they refer to regions or countries, is the multiplier effect, that is the indirect results that can ensue from a particular investment. There is not only the question of what a company will buy locally, but also what its employees will spend. Unfortunately, however, the same investment can have quite different consequences, depending on the company's policy in other respects. For instance, it appears that the huge investments that the electronics industry has made in Scotland have not had anything like as much positive effect as expected, because the various companies have only located their production there, but have continued to buy elsewhere. Similar results have been reported from the gas operations outside Lancaster.[3]

A related topic concerns the dissemination of technology. The spread of innovation is a special case of the anthropological study of cultural dissemination processes. The classical innovation study in the management field (Rogers 1962), for example, is largely based on anthropological studies of this kind. These studies have concentrated on the level of the individual and on the dissemination of general aspects. Much less has been written about the spread of technology in the corporate world. Czepiel (1974), von Hippel (1978), Granstrand (1979), and Takeuchi and Nonaka (1986) are among the exceptions, but we still know relatively little about how technology spreads between business companies.

In connection with regional development and the dissemination of technology, the role of certain special knowledge units has become a vital issue. Science parks, for example, have evolved because the importance of universities and technological institutes as major centres of resources has been growing increasingly obvious. A good deal of the very lively interest which the universities have attracted can be explained by the reputation of Silicon Valley and some other regional areas with a high-technology profile, but the existence of areas with a special technological character goes back much further than this. Ever since the early days of industrialization, special techniques – and in many cases specialized products – have been developed in certain areas. In Sweden there was a shoe town, a locksmith town, and so on. The idea of such centres has therefore always attracted politicians and innumerable attempts have been made to create them; most have failed. Science parks are simply a natural development of this tradition, and some researchers are already warning us that the effects will be far less than we expect (McDonald 1987). However this may be, the question of the role of the universities remains an interesting one. What contact do companies generally have with research institutions?

Even more interest has been devoted to national programmes for technological development. The most famous of these are certainly the Japanese national programmes, which other countries have made various attempts to imitate.[4] The fundamental question is this: by accumulating basic knowledge in key areas and by supporting knowledge development and organizing the necessary complementary technical development, and even by steering technological development itself, can the state help to strengthen the competitive thrust of its corporate sector? The national programmes are the most obvious example, but to many people the major defence and space projects appear equally important. Two vast programmes, both of which are currently the subject of lively debate, are Star Wars in the United States and Eureka in Europe. Sweden can also provide a few examples, albeit on a very much smaller scale, for instance the long-term programme being run by the National Board for Technological Development and investments such as JAS and Telesat (a Swedish communications satellite).[5] And at the regional level the Swedish government required all the county administrations in 1986 to review their own activities as regards the dissemination of technology. The regional development funds represented a first move towards supporting corporate investment in development, and in several counties these are now being supplemented by various kinds of technology centres intended to encourage local corporate technological development, with the help of training schemes or direct collaborative efforts.

However, technological interaction within the corporate world or between this world and other knowledge units is not very well documented. The United States can boast a few surveys, such as Rogers and Larsen (1984) but almost without exception any such documentation is limited to high technology areas.[6] Thus it seems that if we are to be able to support the process more effectively at the level of society, we first need to learn more about what is happening at the corporate level.

The purpose of this book

Decisions-makers in companies and societal institutions are often confronted by situations in which the kind of problems and opportunities we have been identifying represent an important element. Corporate actors in positions of responsibility may be variously well-informed about what is happening in the world outside their company, and they may be capable of analysing this information and drawing the best conclusions from it. And again, they may be open to the idea of exploiting the many opportunities that arise.

My aim in the present book is to describe and analyse technological development in companies, and the associated interactions

Technological development and the company

with other companies and organizations. It is my hope that readers, whether they work at the societal or the corporate level, will

1 learn more of the nature of this interaction and the forms that it takes
2 acquire tools for analysing the interaction and its effects at the corporate and societal levels
3 get some idea of ways in which this type of interaction can be developed further.

In other words the reader of this book should, I hope, become more competent in handling and exploiting the opportunities generated by technological interaction between companies and organizations.

But the book is also the result of an investigation, which in turn constitutes part of a research programme, and it therefore also has certain specific goals connected with this programme.[7] The programme consists mainly of case studies concerned with technological development processes stemming from a variety of contexts. The development of a new paper-pulp production process is one example (Waluszewski 1988; Håkansson and Waluszewski 1986). Another is the development of a new technology field concerned with automated image-processing analysis (Lundgren and Björklund 1988). A third example consists of several studies of technological development within single industries, such as the steel industry (Håkansson 1987; Laage-Hellman 1984, 1987) and the biotechnology field (Laage-Hellman and Axelsson 1986; Laage-Hellman 1988). A fourth is the influence of the purchasing function on technological development (Axelsson 1987). In all of these studies except the last, the company as a clearly defined unit is of minor interest. In fact, one of the main points is that the corporate boundaries are vague and blurred and therefore not suited to an analysis of complete development processes. However, the companies are important as influencing agents in relation to different development processes. Therefore I will start here from the individual company and examine its opportunities for acting in technological contexts within the broader context of its environmental interactions as a whole. In this dimension the present book thus complements the other studies in the wider research programme.

Focus on forms of collaboration and developmental relations

As our introductory discussion has shown, a company's overall interaction with its environment contains so many dimensions and aspects that it is impossible to study it in its entirety. I have therefore concentrated in the first instance on what I regard as the core of this whole interactive situation, namely the collaborative forms adopted by the company in its developmental relations with other units. By a developmental relation I mean any link with an external unit which contains an identifiable element of technological development. It is not necessary that the link be formalized: any relationships with customers, suppliers, or other units that involve some sort of technological development activities are included.

My main reason for focusing on developmental relations and collaborative forms (the way in which different developmental relations are combined) is that the most important way in which a company can exploit the competence, knowledge, or technical plant of another unit in a more advanced or meaningful way is, in my opinion, through developmental relationships. This view will be discussed in further detail below.

One consequence of this self-imposed constraint is that it excludes any interaction between companies and other external units consisting of activities such as general technological monitoring or the recruitment of technological personnel. One could of course assume that these general forms of interaction are related to the specific collaborative mode adopted with individual parties, but their inclusion does not seem necessary to the present study.

Content of the book

The book is divided into three main parts, each subdivided into several chapters. The first part, Chapters 1–4, introduces various premisses – some problem-oriented and some theoretical – of importance to the study. The second part, consisting of Chapters 5–8, reviews the results of an empirical study. The last part, Chapters 9–11, consists of reflections on the material and a consideration of its implications. The following is a brief summary of the content of the separate chapters.

Chapters 2 and 3 provide a survey of the theoretical frame of reference. In Chapter 2 the individual company is studied in its role as part of a network. The network is defined in terms of the actors (the companies) included in it, their activities, and the resources used or processed in the course of these activities. Certain relations between actors, activities, and resources are identified, with special reference to links with other companies. These links are of course a resource in themselves, but they are also used as a means of controlling other resources. They are activities in themselves, but are also used as a way of tying other activities together; and while they

Technological development and the company

are of course created by actors, they also bring actors together in networks.

Chapter 3 discusses the development process in networks. It is argued that the technological dimension represents an integral part of the network, and cannot therefore be separated and analysed on its own. Instead we must consider technological development in relation to the development of the network as a whole, and the individual company's technical contribution in relation not only to this overall development but also to the way in which its own position in the network can change.

Chapter 4 describes the design of the study which provides the empirical basis for the book. This study consisted of interviews and other material collected from 130 small or medium-sized companies or corporate units. The interviews were comprehensive and penetrating; altogether about 300 dimensions referring to action, resources, and corporate characteristics have been measured. The size and dispersion of the corporate sample was such as to allow – statistically speaking – relatively certain conclusions.

Chapters 5–8 contain an account of the results, and a theoretical exposition of several related aspects of corporate technological development. The company's overall network activity is discussed in Chapter 5. The corporate identity, defined here primarily in terms of the company's predisposition to co-operation, is the fundamental premiss. The chapter includes an analysis of the element in corporate technological development that involves collaboration, and discusses the corporate 'co-operation profile' in terms of the intensity of the co-operation and the numbers and types of other parties involved. Furthermore, the effect of the co-operative predisposition on profit and growth in volume is analysed.

Chapter 6 analyses the effect of a company's resource structure on its network activity. Five areas of resources are discussed: marketing, input goods, capital, personnel, and technology. Each one is described and analysed with reference to direct corporate control through some form of ownership or to control by way of diverse links with other parties. In both cases structure is related to the company's network activity, particularly in terms of its predisposition to co-operation.

Chapter 7 is devoted to the relationship between certain basic activities in the company and its network behaviour. Three types of basic activity are distinguished: the company's overall development activities, the production activities that dominate the company's technical core, and the organized activities that unite the company and link it to various other parties.

Individual developmental relationships are described and analysed

Technological development and the company

in Chapter 8. This time we start from certain characteristics of other parties and look at these in relation to the studied company, while also identifying the forms of co-operation that are adopted. The extent of the personal contacts in these relations is also discussed, as well as the results achieved or expected.

The third part of the book, Chapters 9-11, presents some of the accumulated reflections arising from the study, and certain implications for various types of decision-making. Chapter 9 provides a final summary under four main headings. The first is the strategic impact of the relationships and the technological collaboration on individual actors (companies) and on the network as a whole. The second is closely related to this in that it is concerned with the investment aspect of the relationships and the collaborative arrangements. Relations that have taken a long time and a lot of hard and costly work to build up represent what is perhaps the most important base for technological co-operation. The third heading thus concerns the types of parties with which a company can co-operate, and how the different types can be combined. Finally, the fourth point concerns the distinctive characteristics of these relationships, and how they may affect the individual company. Chapter 9 closes with a section on the theoretical implications of the study.

Chapters 10 and 11 are devoted to a discussion of the broader implications for business managers and decision-makers concerned with industrial policy. In both chapters the empirical base is complemented by other studies and consultancy projects, which we have conducted and in which we used the network approach. Chapter 10 focuses on business implications and is divided into five sections, each related to one basic management issue. The first concerns the identity of the company within the network structure. This identity is to a large extent built up from the resource base of the company, which is the second issue. The third concerns monitoring the network, and the fourth the design of the company's co-operation profile. In the fifth and last section organizing is discussed as a general means in the networking process.

The implications for industrial policy-makers are examined in Chapter 11. The discussion is structured around three projects focusing on different basic policy issues. The first concerns the development of a whole industry or a whole technological field in a single country. It is exemplified by a study of investments in biotechnological research and development (R&D) in Sweden, and includes an analysis of the interaction between university departments and private companies. The second issue concerns the general interaction between university research and technological

13

development in companies. The study which exemplifies this issue was aimed at identifying the connections between several major research programmes and individual industrial companies. The analysis included all the different ways that companies can influence and can be influenced by research programmes. The third and last issue concerns development within a broader technology but in a geographically more limited area. The investigation in question examined technical development in and between mechanical engineering companies within one region, in order to see how far they could be regarded as a network and to evaluate what means were available to the local regional bodies for stimulating the development of a local network.

Notes

1 To be in such a room (in England called 'locomotions') gives an immediate sense of the difficulty in evaluating the direction of movement in both the vertical and the horizontal dimensions when all reference points are also moving.
2 The models are mainly designed to give the optimal solution to each separate situation.
3 The experience from Lancaster has been analysed in network terms in Smith and Easton (1986).
4 The experience of the Japanese programmes is discussed in, for example, Okimoto *et al.* (1984), Tung (1986), Patrick and Meissner (1986), and Sigurdson (1986). The Japanese company as part of a network of companies and governmental bodies is the subject of an excellent analysis in Yoshino and Lifson (1986).
5 An overview of the Swedish way of handling these questions can be found in STU-Perspektiv (1986). STU is the Swedish National Board for Technical Development, and is discussed further in Chapter 11.
6 Some good examples are Okimoto *et al.* (1984), Malerba (1985), Blumenthal *et al.* (1986), Dibner (1986), and Peters (1987).
7 The programme is financed by grants from the Swedish National Board for Technical Development and the Marketing Technology Centre. The research is conducted by two university departments in co-operation: the D-section at the Stockholm School of Economics and the Department of Business, University of Uppsala.

Chapter two

The company in a network

A key premiss in the present analysis is that each individual company enjoys important links with other units in its environment. This applies from the very moment of its birth, since many production companies have their roots in existing companies or other organizations.[1] As time passes, various changes occur as the company distances itself from some parties and draws closer to others. Even the closure or liquidation of a company bears the stamp of such relationships, which generally embrace all or part of the discontinued unit. Throughout its life the company is thus marked by the fact that it is not a free or independent unit; on the contrary, it represents part of a network – and, what is more, a network with which it is not even fully acquainted.

Within this network the company tries more or less systematically to influence and exploit various events to its own best advantage. The company participates actively in a stream of events together with other organizations. These events are partly determined by the existing nature of the network, but they themselves also have an impact on the network. Without the support of at least some of the other participating units it would not be possible to make any changes; with support, almost any change becomes feasible. Thus the network embraces a variety of 'political' processes whereby the various actors seek support for their own proposals. The effect on the network will depend on the support that a proposal receives. Some proposals acquire considerable support in that some of the companies involved accept them and start adapting to them, so that through this they begin to have an impact on the network. Others receive less marked support and therefore make only a minor impact, while others receive no support at all and leave no trace. The activities that individual companies perform within the network will thus be affected by these interactions and will be interlinked in a variety of ways.[2]

The network could be described as the arena in which the

The company in a network

company operates. And yet it is much more than an arena: it is also one of the most important instruments of change, since corporate relationships can be used for mobilization purposes. Each company is unique in terms of the links it causes to be forged. Various kinds of external parties may be involved: customers represent an important group, as do suppliers. Together with the company these groups can be said to form a vertical structure. There is also a horizontal structure, consisting of competitors and companies producing complementary products or services. When it comes to technological development, the various knowledge-specializing organizations constitute another important category. Typical of the network is that it links certain specific units together in a way that is unique.

To be able to fulfil a particular role in the network, the company must possess some special capability. In an industrial network this capability is associated with various types of resources. To be able to produce a particular product, it is necessary that the company should have access to raw materials or that it should be able to buy raw materials or components from other companies. Further, it is necessary for the company to have access to appropriate equipment or personnel with experience of the appropriate theoretical knowledge. Finally, access is required to a sales network. And to be able to acquire these resources, some source of financial capital is also necessary. Thus the network links together not only different types of units and operations but also different types of resources.

We can now proceed to a fuller description of the network as a theoretical analytical tool, by defining the primary elements identified above and the interdependencies between them. Furthermore, we can look more closely at the strands that actually constitute the network, namely the corporate relationships.

Theoretical model of industrial networks

An industrial network consists of companies linked together by the fact that they either produce or use complementary or competitive products. Consequently the network always contains an element of both co-operation and conflict. To provide a theoretical basis for describing networks, we can consider three factors: actors, activities, and resources. An industrial network thus consists in theory of actors linked together by their performance of complementary or competitive industrial activities, which implies that certain resources are processed as a result of other resources being consumed.[3] Each one of the three components – actors, activities, and resources – is dependent on the other two. Actors are defined by their performance of activities and their control over resources. Activities are

The company in a network

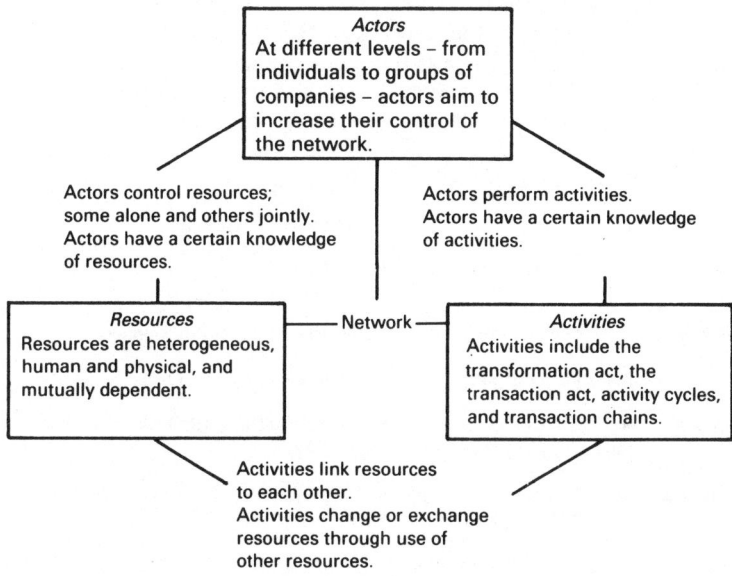

Source: Håkansson (1987: 17)

Figure 2.1 Network model

performed by actors, a process during which certain resources are used in order that others should be refined. And, finally, resources are controlled by actors and their value is determined by the activity in which they are used. Figure 2.1 shows some of the most important links between the three components. They will each be described in greater detail below.

Resources

Resources represent a necessary condition for all industrial activities.[4] We need only consider the example of oil to recognize the great impact that a single raw material can have on the design of various industrial activities, and so to understand the importance of controlling resources and all the actions that are undertaken to acquire such control.[5] In the case of the individual company five types of resources can be identified, each related to some part of the corporate environment. They are input goods, financial capital, technology, personnel, and marketing.[6] The importance of controlling any one of these will vary at different times, as will the individual company's opportunities for acquiring such control.[7] This control can be exercised in two ways: either directly in that the company formally owns the resource or the right to use it (hierarchic control); or indirectly, in that the company has a close and stable relationship with a unit which possesses formal control over the resources. Table 2.1 shows examples of both kinds of control.

Table 2.1 Resource control

Type of resources	Type of control	
	Direct (hierarchic)	Indirect (through another unit)
Input goods	Own the supplier	Stable relationships with suppliers
Financial capital	Own the bank	Stable relationships with banks
Personnel	Slaves	Stable relationships with blue- and white-collar workers
Technology	Machines, production plant	Stable relationships with customer/supplier/research institutes
Marketing	Patents, know-how	Stable relationships with customers

A feature typical of all resources is that the value or utility of one resource always depends on the resources with which it is combined. Thus in economic terms resources are not homogeneous; for this it would be necessary for the value of a resource to be independent of the other resources with which it was combined. Rather, in this sense, resources are heterogeneous.[8] In the case of human resources this is an obvious fact, and one that has been regarded by economists of a micro-theoretical bent as one of the fundamental causes of the evolution of companies. Briefly, the argument runs that the performance of worker A depends on those with whom he cooperates. In other words, as a resource A is not a constant; he is affected by his co-operation with others. For such combinations to be effective, someone has to learn to know the different people or resources available, and thus to discover the right combinations. The company or organization is better at dealing with this aspect than the market; and so, it is concluded, heterogeneity is one of the factors behind the evolution of the company.[9]

At the same time we know that the same effect applies in the field of technological development. A new combination of two resources can generate interesting new products, which in turn affect the value of the resources. Or an old solution may be applied to a new problem

The company in a network

so that new uses for old resources are created, and so on. Thus the value of one resource is determined by the other resources with which it is combined in various activities. All of this means that knowledge and learning about resources are also important functions.

Activities

The activities carried out by individual actors do not occur in isolation; in most cases they can be seen as links in longer chains. Figure 2.2 shows the way in which activities performed within and between different companies are linked together. The activities performed within the company will be referred to henceforth as production activities, and those occurring between companies as exchange activities or simply exchanges.[10]

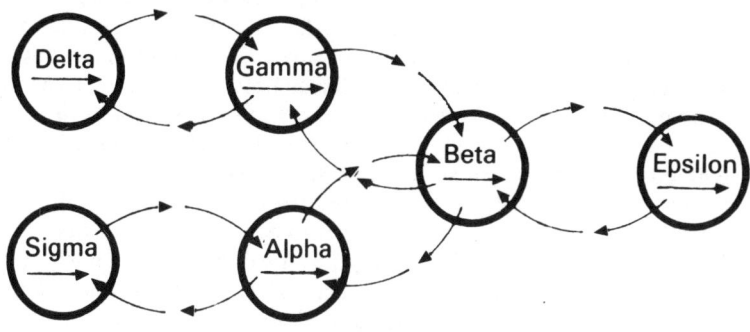

Figure 2.2 A network of six actors

Figure 2.2 shows that the various exchanges and production activities depend on other exchanges and production activities having taken place first, and that they then provide the basis for further exchanges and activities. These dependent relationships are important in the perspective of technological development. A technical change in one relation, perhaps between Alpha and Beta, can affect all other activities. If the change means that the product which Alpha sells to Beta has been altered in some basic way, then Alpha may have to make new demands on his supplier, Sigma. The change may mean that the complementary product which Beta buys from Gamma also has to be changed, which in turn may affect Gamma's relationship with Delta. The change could also mean that Beta can offer Epsilon a better product. A single change in one activity always has some effect, great or small, on other activities in the network.

This description demonstrates that there is a considerable degree of co-ordination in the exchange activities. In an industrial network this occurs in several dimensions such as time, technical function, form of supply, and so on. The co-ordination is so complex that the actors often have to enter into close relationships with one another, and within these relationships co-ordination issues are dealt with on a continuous basis.

The need for co-ordination is determined by the dependency relationships between the activities. Several types of dependencies generally co-exist, as the above example demonstrates. There can be sequential dependencies, whereby some activities must be carried out before others. At the same time there are shared dependencies, whereby two or more activities may be mutually dependent because they are all linked to some other single common activity. Some activities are mutually dependent, perhaps because the results they achieve are later going to be used together. If one of the activities in such a case is changed, the other will also have to be altered to avoid problems later on.[11] These various forms of dependence stem from the fact that the activities all perform a function within a larger whole. Each activity must be adapted to this greater entity; or we could say that if every activity is adapted to the greater whole, this overall function will be better performed. If every activity only belonged to one larger whole, adaptation would be no problem, but difficulties arise when an activity belongs to several wholes at the same time. The design of the single activity then represents a compromise between the demands of the different wholes on the one hand and the demands deriving from the internal nature of the activity itself on the other. Unfortunately, every production activity that is separated by activities from other parts of the whole is certain to belong to several wholes. This follows more or less by definition; if there is to be any meaning in identifying a particular production activity and relating it to others by way of exchanges, then the activity will have to be included in several such wholes. Its design is therefore a compromise – and what is more, one that will always be questioned; the relationships (exchange activities) exist partly to bridge the gaps between production activities and partly to channel and handle a variety of conflicting forces.

Actors

A major characteristic of the actor element in industrial networks is the great range of variety that exists. Individual people can be important actors, or groups of people, or departments in a company, whole companies, or even groups of companies. In fact in certain contexts – and technological development is an example – the very existence of this variety is one of the most important characteristics (Waluszewski 1988). In the present study, however, we are concentrating mainly on the corporate level, and can thus restrict ourselves to actors in terms of companies.[12]

As we have seen, every company performs certain activities and controls certain resources, which in turn implies that it has certain specific links with other units. In addition, it is distinguished by certain specific interests of its own. The company thus possesses a particular identity in the network, and in addition a certain amount of power. But the actors also bring a dynamic element into the network, since they are prepared and able to generate change, which they do by altering activities or resource combinations either directly through their own actions or indirectly by causing others to make changes as a result of their mutual links. An important part of these changes can be defined as technological development. The actions of the individual company are based on the company's identity in the network (in terms of the existing activity and resource structure), as well as on its knowledge and its desire to create new structures.

There are two very special qualities attaching to these actors (the companies). The first is that the particular combination of resources, activities, and other parties in the network makes every actor unique. Every actor therefore also faces unique demands and opportunities. Theoretically every actor enjoys an infinite number of potential development opportunities, but of these only a few are ever identified; even fewer are seriously considered, and only a fraction are tried. Any chosen line of development, which is consequently also unique, generally consists of many small steps taken over a long period, during which the company successively adapts itself in a lengthy and laborious process to the changes occurring in its environment. Occasionally an attempt may be made at some more radical reorientation, which may be registered primarily as a change in the priority assigned to important co-actors. Some of these attempts do lead to major changes, but many are mere ripples on the surface which quickly disappear. In this way the uniqueness of the actor is evolved and maintained.

The other special characteristic of the company/actor is its resource-dependence: the ties represented by the investments it has made in plant, personnel, and, not least important, in its links with other parties. The actor's existence in the network is totally dependent on these investments; at the same time, however, the company is their captive. Most activity changes call for changes in resource structure, and that means costs. Thus in face of any change it is always first a question of how existing resources could be utilized differently. To start from scratch in a completely new field where existing investment cannot be used, for example, is only likely to be interesting in the most extreme situations.

Corporate relationships

One way of controlling resources, as we have seen, is to enter into a close relationship with a unit which has control over certain resources. At the same time we have declared that one of the most important resources that a company possesses consists of the relationships which it has built up with various other units in its environment. In discussing activities we emphasized the need for co-ordination, and relationships naturally represent a crucial way of co-ordinating production activities. Relationships involve exchanges, and therefore in themselves represent activities.

Companies forge links with one another for reasons connected with resources and activities. There may be a mutual interest in the link, or its value may be more or less one-sided. An interest in forming a relationship may also spring from some link with another party; it is easy to ignore conflicting interests if there is a greater threat from a third direction. Thus one company's perception of its own interests in relation to various other parties is relative and situation-dependent. Its relationships therefore always contain elements of harmony and conflict, of mutual and contradictory interests.[13] If the element of reciprocity dominates, a partnership may evolve; but if conflict dominates, the relationship is likely to be short and limited in nature – if indeed any relationship is born at all. When strong mutual interests and sources of conflict exist side by side, the relationship will tend to fluctuate over time, depending on which interest is the more relevant at any particular moment. Thus an important point about these relationships is that they can contain a substantial element of conflict and yet still work, for instance when it comes to co-ordinating technological development.

We can now examine in greater detail the nature, content, and function of these relationships as they affect the individual company.

Characteristic features

Various earlier studies (Håkansson 1982; Turnbull and Valla 1985) have told us a good deal about individual relationships. Five significant and well-documented characteristics refer to the duration of the relationships, the adaptations involved, the technological content, the range of contact, and the social content.

Typically, many of the more important relationships are well established, often having lasted for several decades. This means that the two parties know each other very well, and this in turn affects their behaviour. In such a long-lasting relationship the same actors will have met one another countless times; both sides will recognize that the wrong handling of a particular situation could spoil the whole relationship and waste all the sacrifices that have gone into building it up.[14] The continuous contact also means that both sides have learned how to handle variations in their relative positions, perhaps when the economic climate affects them differently or at different times. Thus the relationship contains a dynamic element; at any single moment it can be regarded as a compromise, applying just then but ready to adjust to anything that happens inside or outside the relationship.

In many of the most important relationships all kinds of adaptations and adjustments occur. It may be a question of products or production processes, methods of supply, delivery times, financial arrangements, or some adaptation on the knowledge front. Adaptations create certain ties but they also promise big gains in efficiency; cross-company arrangements generally have their advantages, and individual adaptations are often a result of such considerations. Sometimes, of course, they may result from an unequal balance of power, so that a large buyer has perhaps dictated the conditions of an agreement with a smaller subcontractor.

The third characteristic feature is perhaps the most interesting in the context of the present book. Relationships often prove to have quite a substantial technological content, particularly when the parties involved are production companies employing heavy technical plant. In general terms we could say that such relationships are intended to link two technical resource units together, in such a way as to maintain the efficiency of day-to-day operations at a high level and to sustain full strength in development capacity. Since technical issues represent such a crucial part of these relationships, the whole issue deserves a more thorough analysis.

The multifaceted nature of these relationships is also reflected in the many decision-makers involved on the various sides, decision-makers who represent different functions or departments of their companies. Production and development personnel, for example, are often included. There is often frequent contact, perhaps at weekly or monthly meetings. Thus each such relationship should be regarded not only as single link or strand, but often as an accumulation of links or strands involving a number of people. In this perspective it can be seen that the cost of maintaining the relationships is likely to be fairly substantial.

The fifth and last characteristic feature refers to the social content of the relationship. According to Blau (1964; 454), social exchange implies that confidence is gradually built up between two individuals in a long process during which the parties try one another out. The same seems to apply to corporate relationships. In other words confidence and trust are important but they must be matched with the situation and applied with realism. The parties learn to trust one another in certain respects and within certain limits. They learn to use one another as partners in certain specific ways.

Relationships and commitments

Relationships can also be analysed in terms of the type of bond which they create between the parties. To some extent an analysis along these lines overlaps with our description of the characteristic features of such relationships, but can also serve to complement it in an interesting way. In an earlier study (Hammarkvist *et al.* 1982) a distinction was made between five types of bonds: technical, time-related, knowledge-related, social, and economic/legal.

Technological ties are formed when different parties adapt to one another in some technical way. It may be that a buyer company bases its production apparatus on the purchase of a particular material or component from a supplier, or conversely, that the supplier adapts his products to the final product or the production arrangements of the buyer.

There is usually a strong time-related bond between the parties, which has become more important as a result of a desire to tie up less capital (just-in-time systems). Various administrative routines have thus been brought in line with one another in these relationships in order to handle issues of this kind, and this in turn becomes a bond of its own.

Knowledge-related ties are formed when two parties gradually acquire knowledge about one another, which means that up to a point they are integrating their total store of knowledge. This type of bond has increased in importance as it has become more expensive to keep up with the frontiers of knowledge, and cross-border link-ups between different areas of knowledge have become more common. For a single actor to have any chance of keeping up, suitable knowledge partners – specialists in their own field – are more necessary than ever.

The social exchanges described in the previous section naturally also involve social ties. Confidence and trust imply responsibility and the fulfilment of obligations. Social exchanges impose certain restrictions, mainly in the shape of mutual expectations, and these social bonds are often of undoubted and crucial importance to the functioning of the relationships. We can envisage the relationships as channels for resources. Before the resources can start to flow along the channels, diverse obstacles have to be removed and the channels opened up. One major obstacle is the uncertainty that the parties feel about one another; as a rule people are frightened of trusting somebody they do not know. But the social ties are more than just a way of removing obstacles; they also create better opportunities for actively influencing the environment.

The fifth and last type of bond consists of agreements and legal ties, in other words the different forms of owner influence, joint ventures, and so on that have been becoming increasingly popular. The value of the legal agreement used to be regarded largely as a question of insurance, as something to fall back on if things went wrong. Certainly this has been one of its functions, but another is to make the bond visible to others. Sometimes visibility may even be a goal in itself, since companies, like people, are often judged by their friends.[15]

All in all, such relationships represent a constraint on independence and freedom; the company is bound to certain other parties and thus to a structure, a network. But the same bonds provide an opportunity for influencing other parties which have also made commitments. In this way bonds and ties become an instrument and a medium for interactions and counteractions between a company and other units in its world.

Functions of the relationships

We have described the various relationships in terms of their external qualities and their importance simply as relationships. The next stage is to describe their function with reference to the individual company. The relationships are important in at least three respects: they affect the company's productivity, they affect its control over the environment, and they affect its developmental strength.[16]

Relationships affect productivity in that the technological design of a company's production activities is related to that of its suppliers and customers. Relevant examples are just-in-time deliveries and integrated ordering and invoicing systems. The relationship links the company's technology to the environment, and the efficiency of the link will determine the company's performance, that is, its productivity.

The relationships are also important from the point of view of control and stability. They may link the company with the environment, but they also tie the environment in with the company.

It becomes easier for the company to acquire early warning of change, and even to ward off or mitigate change of too violent a kind. All in all, the company has more control over developments.

The relationships are also important when it comes to technological development. The company has more opportunity to find out what is going on technologically; it can take part actively in various projects conducted by other parties and it can even encourage other parties to develop technically in such a way as to favour its own interests. In the present study, we shall be concentrating on this third aspect, but in order to study it in depth, we shall also have to consider the other two.

Summary of the theoretical framework

The company represents part of a network. It has relationships with a number of other units and the relationships link these units together in a network structure. The relationships exist to handle various kinds of interdependencies generated by the activity and resource structure. The company acts within the framework given by the network of its relationships. Its actions are marked by both conflict and harmony because of the simultaneous presence of conflicting and common interests.

The technological dimension is evident in the actions and structures that create the dependent relationships. Technological dependence occurs between different activities, while technology also ties different resources together. If there is to be a change in resource use, then technical changes are needed, just as they are needed to alter the activity structure. The individual company's role *vis-à-vis* technical changes can assume several forms: to some, it adjusts; others, it initiates or supports in various ways; yet others, it opposes. An important element in its technological behaviour consists of collaborative projects with various other parties. Before proceeding to a more detailed analysis of such collaboration, let us look a little more closely at development problems in networks as a whole.

Notes

1. In a study of fast-growing technology-based companies in Sweden the origin of these were investigated. The normal case was that the entrepreneur(s) earlier had worked in a large company and furthermore that this company was often heavily involved in the development of the new company as a main customer or supplier but also often as a financial source (Utterback and Reitberger 1982).
2. This duality of the network – that it functions as a constraining factor when it is not mobilized but as a powerful changing force when it is –

cannot be stressed too much as it is the mechanism that all the actors try to use in their favour.
3   This theoretical model has earlier been described in Håkansson and Johanson (1985; 1988) and in Håkansson (1987). It has been developed in an inductive way through empirical studies of industrial markets. Important theoretical inputs into this process have been given by Lindblom (1959) and Cyert and March (1963) regarding decision-making within companies; by Thompson (1967), Weick (1969; 1976), Van de Ven et al. (1975), and Scott (1982) regarding the way companies are related to their environments; by Alderson (1965), Richardson (1972), and Williamson (1975; 1979) regarding how collectives of companies function. Recently published works giving similar or closely related views are Lorenzoni and Ornati (1988), Powell (1987), Axelrod (1985), Smith and Easton (1986), Kagono et al. (1985), Takeuchi and Nonaka (1986), Johanson and Mattson (1985; 1988), Mattson (1985), Melin (1983), Eccles (1981), and Kinch (in press).
4   Our network approach is in this way closely related to the resource-dependence approach as given by Thompson (1967), Jacobs (1974), and Pfeffer and Salancik (1978).
5   A good description (but in different terms) of how the network developed within the oil industry is given in Sampson (1976).
6   This way of grouping resources important to the company is taken from Axelsson and Håkansson (1979). A similar grouping is given in Jacobs (1974).
7   As an example from Axelsson and Håkansson (1979), the Swedish steel companies in the nineteenth and early twentieth centuries had to control the supply of wood (as charcoal was an important input) and iron ore as these two resources together determined the quality of the end product. Neither of these resources is of any major importance today due to the development of technology.
8   Heterogeneity of resources is one of the basic assumptions behind the network approach. For a more complete discussion, see Hägg and Johanson (1982) and Håkansson (1988).
9   This theoretical theme is adapted from Alchian and Demsetz (1972) and is often referred to as the explanation for the existence of companies in the theoretical world of markets.
10 The chain of activities described here is different but still comparable with the value-chain discussed by Porter (1985) and others, as well as the distribution activities within distribution channels as discussed by Mattsson (1969). See Gadde and Håkansson (1988) for an analysis of distribution problems in network terms.
11 For a further discussion see Ford et al. (1986). The grouping of different dependencies is taken from Thompson (1967).
12 In the section about personnel in Chapter 6 some further comments regarding this limitation are given. An interesting case study showing the importance of identifying different 'levels' of the actor is given in Liljegren (1988).
13 A more comprehensive analysis of the role of the relationships is given in Håkansson (1987: 8–13).

14 The development of co-operation within these relationships follows very much the same logic as described by Axelrod (1984).
15 The difference between informal and formal co-operation projects and relationships is discussed in detail in Håkansson and Johanson (1988).
16 For a further discussion see Hägg and Johanson (1982: 77ff) and Håkansson (1987: 10ff).

Chapter three

Technological development in industrial networks

The technological dimension is an integral part of the network, not to be separated and analysed in isolation but to be studied in association with the development of the network as a whole.[1] We will start by analysing the technological development process in a network in relation to certain other general network processes: first, the process whereby actors combine activities and resources, and second, the process whereby actors struggle for control over resources and, consequently, over activities. Technological development is relevant in relation to both these general processes.

In the second part of the chapter technological development will be analysed from the point of view of the individual company, and again in relation to the overall development of the company as whole. Three general processes, in which technological development represents an important ingredient, are identified.[2] The first is connected with the way in which the company behaves towards other units in its network. Within its framework of interactive behaviour it can integrate its operations to a greater or lesser extent with those of other units. For instance it may establish several very close relationships or it may keep a certain distance. A natural integrated ingredient in this situation is its behaviour in connection with technological development – whether, for example, it engages in collaborative projects.

The second process is concerned with the handling and developing of the company's resource base. Resources are important on two counts: they affect the company's overall power position, and they represent its capability to perform various activities. Two aspects therefore have to be considered, namely how the resource base should be changed over time and how it should be exploited in various activities. Both aspects confront the company with a number of questions, problems, and opportunities, which we have chosen to identify as a single all-embracing process. Technological development is closely connected with both aspects, and thus clearly represents part of this process.

Corporate development as a whole is the basis of the third process. A company's overall development seldom proceeds at a calm and even pace, and varies from one period to another. Sometimes there may be a great leap, a break with the past, while in between there are longer periods of gradual change. Here, too, technological development is a crucial part of the picture.

The development process in networks

The industrial network changes continuously and in a variety of ways. Activities and resources are combined in new ways, as activities evolve to make better use of resources and new knowledge of resources opens up new areas of use. The actors drive this process in order to acquire advantages for themselves, but they are also looking for ways of increasing their control over critical resources. Technical development is part of this, and itself represents a process whereby new technical solutions are designed, tested, changed, rejected, or accepted.[3] The process is political in that all those involved are trying to pursue the line most advantageous to themselves and every actor thus hopes to drive the process in a particular direction; but it also contains a certain random element due to the presence of genuine uncertainty.[4] This means that knowledge development can assume either theoretical or practical form. But companies cannot always tell whether it will be possible to solve certain technical problems, given the prevailing economic constraints, or whether certain theoretical solutions will prove feasible in practice.

Thus the technological development process in networks posesses major elements of conscious will and of unpredictable uncertainty. The process is by no means predetermined nor can the outcomes be predicted, but certain patterns in the way development proceeds can none the less be distinguished through its connection with the overall development of the network.

Over periods of any length the typical pattern for an industrial network seems to be one of gradual development interrupted by shorter periods of more dramatic change (Håkansson 1988). During these latter periods some fundamental principle or principles that had steered events during earlier development periods are changed. In the course of the transformation the established structure of the network is challenged, while the conditions of the new structure are being created. During the next stage of development this structure is gradually honed until it is once again time for a transformation. Obviously the intervals between transformation periods can vary enormously, partly depending on the type of dominating technology in the network.

It seems that underlying causes of this alternation between transformation and development spring from an interaction between two basic network processes: one concerns the way in which actors combine resources and activities and is defined by concepts such as 'structuring' and 'heterogenizing', while the second concerns the actors' efforts to strengthen their control over resources and activities and is defined in terms of 'hierarchization' and 'externalization'. These processes will now be described in greater detail.

Combining activities and resources

Industrial networks are based on a particular combination of activities and resources; their structures always reflect the way the combination is made. The combination becomes 'set' as a result of all the investments in plant, knowledge, and relationships made by the actors involved. Major investments automatically represent fixed points; they determine activities, and this then affects relations between companies. They also decide resource use. Single big investments in different parts of the network provide the general framework for other activities. At the same time many investments evolve gradually in a long series of continuous activities. Some are visible and obvious to all, such as plant. Others, such as relationships, language rules, and so on are much more difficult to observe, but can none the less be more important than the obviously visible ones. Some are undertaken separately by individuals, others by several companies together. Since every investment in itself implies a commitment, companies generally recognize that any such new ties must accord with what is happening elsewhere. The result is a continual process of adaptation. In addition to this, investments undertaken together with others always mean that two parties must adapt to one another in some way. Altogether the accumulation of adaptations in the network means that the companies involved will function better, but it also makes it more difficult for them to co-operate with companies outside the network.

From an examination of the effects of all these changes, whereby some companies adapt to one another and consequently lose the chance to work with certain other companies, a development pattern seems to emerge. At least in part this can be defined with the help of the concepts of structuring and heterogenizing.

Structuring implies that a series of investments are made individually or jointly, whose effect is to reinforce the established structure. These investments build further on the fundamental combinatory principles on which the existing network is based. All investment in rationalizations and large-scale operations is generally of this type, as are many of the small successive changes that can bring a series of activities into line with the basic principles that obtain.

Structuring thus implies that the companies in the network take various steps to get more out of the basic technical idea on which the composition and function of the network is based. Improvements are realized either by refining certain technical relations as a result of successive adjustments between subcomponents, or by exploiting a better quality of materials and by rationalizing and making better use of scale advantages. This last in turn requires a standardization of materials and a scaling-up of production technology. Both methods of improvement involve better use of known resource dimensions. The network is developed and refined along the basic technical lines already laid down. The structure becomes clearer; 'structuring' has occurred.

At the same time and parallel with these structured changes, various other changes which break with the established pattern are taking place in what we can call a heterogenizing process. This is a process of renewal, whereby actors try to exploit new resource dimensions or use known dimensions in a different way. The impulse towards heterogenization comes from human qualities such as creativity and innovativeness as well as from the multifaceted potential of the resources themselves. No resources are fully exploited; there is always a chance of finding new and better areas of use.[5]

Heterogenization complements structuring, but it also represents a threat to the established structure. It acts as a complement in so far as it can generate solutions in the vague areas not covered by the coarser weave of the basic structure. In this way heterogenization creates special solutions, which although breaking with parts of the basic structure none the less continue to support it as a whole. But heterogenization becomes a threat when the very weave is called into question. Structuring and heterogenization are continuous, but sometimes the heterogenization process steps up to such an extent as to challenge the whole established structure, and the stage is set for a transformation of the network. Such transformations generally lead to the departure of some of the former actors from the network and the arrival of new members. Furthermore, there will be shifts in the relationships of the actors who remain.

Here we suggest that a fundamental explanation of increasing heterogenization lies in the actors' struggle for control over resources and activities.

Control over resources and activities

In order to perform various activities, actors need resources. The supply of the various resources is generally limited, and there is therefore a struggle in the network to gain control over them. By directly controlling a resource or by controlling it indirectly through others, a company can augment its own power. Not all resources are equally important from the technical point of view – some are easier to substitute than others – and the supply also varies, which means that certain resources are particularly important to control. But technological changes can affect the relative importance of resources: some may become more crucial while others become less so or may even cease to matter at all.

To acquire control, investment is needed. Consequently companies invest in order to increase their control over resources which they perceive as being important to the future. Since investment requires financial capital, a company may decide at the same time to disinvest with respect to other resources which are thought to be losing in importance. As a rule there will be some common features in the way different actors perceive these things, and this leads to the emergence of certain hierarchization and externalization processes in the network. Certain resources and activities are brought together; others are separated. At the beginning of a hierarchization process, control over the resource is often fragmented and disorganized. Hierarchization implies that the control is organized, becoming successively more controlled in relation to the network concerned.[6] Hierarchization has a natural end point in the monopolization of the particular resource area. When hierarchization approaches this state, those actors who perceive a long-term threat to themselves will sharpen their efforts either to increase their own control or to render control less important. If a great many actors feel control slipping away from them, the forces that will be mobilized to reduce the importance of this control will be correspondingly great. This may be one of the major reasons for the waves of heterogenization that we identified above. If the surge of heterogenization generates a transformation, some of the resources that were previously so important will become less so, and it will no longer be so necessary to control them. The resources are externalized in various ways; they are released and made available for use in other contexts.

Implications for individual companies

The presence of network processes in itself implies an element of collective action at the corporate level. The individual company is no atomistic unit, acting only in its own interests; rather it represents part of a larger whole. What this larger whole actually consists of will change from one situation to another, which is one of the reasons that we use the network analysis to describe the relationship between the company and its environment. Theoretically we could say that a unique network can be identified in relation to every specific problem situation. In other words there are as many networks as there are situations. In many of these situations, however, the networks are very similar in terms of the units included, which in a purely empirical and practical sense is an obvious advantage. Because of the existence of the networks, no individual event, activity, or project can be considered in isolation but must be seen as part of a pattern, as a reaction to earlier events or an invitation to fresh reactions.

Every activity is expected to engender counter-activities, some of which are intended to obstruct it and others to support it. When we try to formulate issues relevant to the individual company in the next section, it is from this complex interaction that we must start.

The corporate development process

The company is included as an active unit in a changing network. It acts (moves) within an area which also moves in relation to other areas. All these areas in turn are composed of other active and mobile units. Any attempt to describe the developments or changes in a unit in the network must take this situation into account; in other words the development process has to be described in relation to others, rather than on its own. Changes in direction have to be identified in relation not only to the overall pattern of change in the network but also to changes in other individual parties.

Characteristic of the technological development process is that the individual company performs countless activities that have a technological content. For example, it engages in a series of contacts during which technical issues are negotiated. Products are changed and the related service is altered or discontinued. Minor changes occur more or less continually, while now and then there will be some major new development. Some of the activities occur internally, others in collaboration with other companies. Production also changes in much the same way, continually in the shape of day-to-day rationalizations, and successively in the shape of major investments in machinery or new plant. Here, too, other organizations can play an important part as suppliers, consultants, or research units. The company's own ambitions and plans affect developments, but so too do the plans and ambitions of various other parties. Sundry chance events also come into the picture.

Thus corporate development is in itself a multifaceted process, and one that also constitutes part of the various processes which we have previously identified in the network as a whole. Let us now try to capture an essential part of this variety by analysing technological development in relation to three other processes, each of which has been identified in terms of the interaction between company and network. The first refers to the handling of relations between the company and other external units. One aspect of the change pattern implies that the company draws closer to some units while distancing itself from others.

Technological development represents a fundamental ingredient in these relationships; conversely, collaboration with other parties should represent an important factor in corporate technical development. According to our theoretical appoach, relationships or links are a characteristic feature of the network, and our empirical consideration of the various issues related to this approach in the next section can therefore be seen as a general examination of the rationality of the approach as a whole.

The second process refers to the base from which any corporate development must start. This base consists of all the resources that the company controls directly or indirectly, including the knowledge on which its technical development is based. We consider here both experience-related and theory-related technological development, and we thus also include the company's different types of collaborative partner.

The third and last process in this context refers to the company's overall development pattern. The company generally moves in much the same way as the rest of the network. Development proceeds in stages, and assumes many different guises. Occasionally, bigger, leap-wise changes occur, generally representing attempts on the part of the company to make a dramatic change in its role and its position in the network. In both sorts of change, by step or by leap, technological development generally plays a crucial part, and we shall therefore be studying this aspect more closely.

Each of the three processes defined above will now be discussed in greater detail, and a number of specific management questions will be noted.

Technological development in corporate relationships

Our discussion of technological development in networks has shown us an abundance of relationships between what goes on in the companies (the actors) and what goes on within the many activities. A natural ingredient in these interconnections is the substantial element of technological co-ordination and collaboration that characterizes a company's relationships with other external units. In a previous context three ways in which advantages can accrue from pursuing technological development in co-operation with other units have been suggested (Håkansson 1987: 92ff).

First, there is an interactive effect in that new knowledge often appears in the border zone between established bodies of knowledge. Interaction with another party that has knowledge of other areas can therefore generate new thinking. Second, a technical novelty that has the support of several companies has more chance of being accepted. In co-operation with others it is easier to mobilize the necessary resources. Third, because of the increase in specialization in production and development at the corporate level, companies need to supplement their own resources with those of others. The first of these points is specific to technological questions, while the other two apply more generally to networks as a whole. Mobilization is connected with corporate groupings *vis-à-vis* other interests, and specialization with an increasingly far-reaching division of labour. In both cases technology is an important factor, but other factors are also involved in the relationship with external parties. Consequently, technological co-operation may often provide the means of achieving goals other than those connected solely with technological development.

Co-operation also has its disadvantages, the first of which concerns control. When a company collaborates with others, it generally loses some control. It has to part with some of its knowledge. At a later stage it may even find it difficult to market the new technique on its own. The company may become the prisoner of its network. Another disadvantage is connected with cost, which rises rapidly as the collaboration becomes more extensive. Cost therefore sets a natural limit to the total number of collaborative undertakings that a company can engage in.

Several questions arise in any consideration of technological development in co-operation with others. How much of a company's total development budget should be allocated to collaborative projects? Obviously there will be considerable variations here. Some companies work in relative isolation, while others do a good deal of their development work in collaboration with others. Up to a point this may be a matter of choice, but to some extent it is probably also connected with the company's role and functional mode in the network. A subcontractor, for example, can be expected to direct much of its development work more or less automatically towards the biggest buyers.

Another question concerns the choice of collaboration partners. It seems very likely, for example, that co-operation on technical development will follow the usual pattern of business exchanges, that is that the first choice falls on customers and suppliers. It presumably calls for more effort to establish contact with parallel units, and it is probably more difficult to create longer-term relationships with such units in the absence of any regular business contacts to which the collaboration can be linked. On the other hand a good many parallel units probably do possess useful capabilities and may well function as a complement rather than a competitor to the company concerned. A company's co-operation profile is thus worth documenting, and it should be possible to relate it to other corporate policies and to the company's position in the network.

A third question concerns the content and form assumed by individual development relationships. For instance, relationships probably vary greatly in their formal shape, whether or not they are expressed in a written agreement, and so on. A formalized relationship becomes more visible, which may be an advantage in some situations but not in others. Previous research has often focused on formalized collaborative efforts, which makes it particularly interesting to look at the frequency and orientation of these, and to see how they differ from more informal kinds of collaboration. Similarly, the content of these efforts – the number of subprojects and their duration – can be expected to vary considerably. Here, too, documentation and analysis could throw light on the situation.

The management issues identified above are important not only in their own immediate context but also in a broader managerial perspective. The first touches on the question of dividing resources between internal development projects and collaborative projects with external partners. Is the present distribution of resources the best for the company, or should a change be made? The second question touches on the order in which other parties are given preference. Does the company exploit its various parties efficiently, or are some being more or less forgotten? Finally, the third question touches on the way in which the company is currently working with the various external units, and whether the potential for variation is being exploited fully.

Development of the resource base

Our analysis suggests that corporate development is affected more by its base and its past than all its future opportunities combined. In other words, it is the base that largely determines which problems and opportunities the company will perceive as relevant and important. And the corporate base in itself can be defined as the combination of resources and activities around which the company has been formed. Technological development represents one aspect of overall corporate development, and has therefore essentially the same base. Essentially, ideas for technological development can arise from two sources: from practice based on the experience of using different technical solutions, including problems previously solved, or from theoretically generated knowledge. In the first case the ideas may emerge as the result of systematic information-gathering or as a flash of genius based on data-collection of a more unconscious kind. In the second case, the ideas emerge as the result of research.

Every company has its own specific development base, consisting partly of its internal resources of personnel and equipment, and partly of its external contacts and co-operative relationships. Every company's development base has certain unique qualities, but there are also likely to be similarities between certain kinds of company.

The relative importance of experience and research as the basis for technological development presumably depends on the kind of technology involved. Developments in biotechnology or automated image-processing, for example, are largely determined by the work of researchers. In engineering and sawmill technology, on the other hand, development based on experience is predominant. This difference affects the way the network functions in a technological perspective, and consequently even the development base of the individual companies. If the research-based element becomes more important, the research units will be more active and dominant and corporate research departments will grow both in size and status. In such situations researchers also come to play an important role in linking the different companies together. The network will then naturally acquire a great many lateral contacts, becoming horizontally ramified.

If the network is dominated instead by experiential technological development, a great deal of the technical dialogue will take place between producers and users, and the network will be vertically ramified. Further, it will be characterized mainly by gradual stepwise changes.

Companies have not only to live in a functional (technical) environment; since they are located in a spatial dimension, they also have a certain geographical environment, perhaps in an industrial centre, a university town, or some smallish community with little industry. In each of these three locations, the immediate environment would be very different in size and in the type of knowledge on offer. Since the company is always embedded in its immediate environment in some way (the staff live there, nearby organizations and companies can act as customers, suppliers, consultants, etc.), these features must affect corporate behaviour *vis-à-vis* external parties. Just as characteristics of the technological field provide one framework for corporate action, various characteristics of the immediate environment provide another.

Three questions arise from these observations. The first concerns the relative importance of the experiential and theory-related bases. Research-oriented units are probably of marginal interest to most of the small and medium-sized companies in the empirical investigation described in Chapter 4. The question then naturally arises as to whether research-oriented units are important at least to a small group of companies and, if so, what characterizes these companies? Are the research units more important to the companies that invest heavily in technological development, for instance? A third question concerns how companies exploit the local knowledge base in their own technological development, and how this varies with the nature of the company and of the immediate environment.

Again, all three questions can immediately be reformulated as management issues. The first then touches on the relative priority given to experiential and theory-related technological development, but it also includes the type of system a company adopts for picking up ideas and suggestions. The second touches on the present importance of different types of research-based units to a company, but includes even their long-term importance depending on the direction taken by corporate development in the future. The third question focuses on the degree to which the immediate environment is exploited, and asks whether the company has the right information about what the immediate environment can offer and whether it exploits this knowledge to the full.

The overall corporate development pattern

During certain periods companies change and develop in a gradual way. These comparatively tranquil periods then give way to more dramatic and eventful passages, when the company's whole direction may be called in question and major changes are made. The changes may take the shape of investment in plant and machinery, rationalizations, personnel changes, training, product development, and organizational changes. These different types of change are intertwined, and combine to form a common development pattern. Technical dimensions – and thus technological development – are a crucial aspect here, and it should be possible to relate them to the two basic types of change: step-by-step (gradual, continuous) and leap-wise (single, discontinuous).

Leap-wise changes may occur in the shape of major investments or radical new products; perhaps a new investment is undertaken, or some kind of production equipment is changed or complemented. This kind of investment affects the company both in the products that it needs and the products that it makes. For instance, new production equipment may mean that the final product has narrower tolerances and a better functional standard, but also that there is less flexibility in the use of raw materials. It often also means an increase in capacity. Leap-wise product changes may occur as the result of scientific discoveries or some other form of radical new thinking, and they can lead in turn to large-scale investment in new production plant.

All companies undergo the kind of continuous technological development that consists of a series of minor technical improvements. These may be on the production side, perhaps making better use of machines or undertaking some complementary investments, or they may consist of minor product improvements or some change in the product range. By sorting through the range and specializing on a more limited number of items, for instance, a company can often achieve considerable scale advantages.

The small successive changes of this kind are generally many and varied. Together, however, they amount to a substantial overall reorientation, normally much greater than all the leap-wise changes together. The gradual changes are also closely connected with day-to-day operations and are often difficult to distinguish from these both internally and in relation to various external parties.

It might seem logical to associate the step-by-step changes with rationalizations (structuring) and the leap-wise changes with new orientations (heterogenization). In practice, however, almost the reverse applies. Obviously, most gradual changes are in the nature of rationalizations, minor product changes, and so on, changes that 'structure' the network. This is probably true of most leap-wise changes too, apart from the major product changes. A large-scale production investment is presumably so inherently uncertain that the company prefers in the first instance to stick to tried and tested solutions. Instead, new orientations can be expected to consist of a number of smaller steps which gradually show that a new path is practicable, before the company feels bold enough to make a major investment in the novelty.

One question that arises from this discussion concerns the connection between the company's collaborative projects with various other parties and the two sorts of change. Most such projects are probably of the step-by-step kind, but at the same time it is probably particularly important from the point of view of the individual company to be able to collaborate with other parties on bigger changes. In such cases collaborative efforts bring the change to the notice of the environment, for example customers and suppliers, and help them to accept it. Similarly – at least if our assumptions about an interactive effect, for example, are correct – such collaboration could lead to even bigger changes. A natural question then arises: is there any difference between the other parties in this respect? A possible hypothesis is that collaboration with customers leads in the first instance to the step-by-step kind of change, while collaboration with partners in the horizontal dimension is more likely to lead to leap-wise changes.

Once again our questions can be reformulated in a management perspective. How should a company best exploit external collaboration to promote leap or step changes? Conversely, how can collaboration be used to ensure that corporate changes become entrenched in the other units in the environment? Which partners can be expected to generate ideas and suggestions for major changes?

Corporate technological development – a summary

In this chapter I have argued that to understand more fully the meaning of the corporate technological development process, we must look at it in a network perspective. The process represents an integral part of the development of the network as a whole, which we have described in terms of structuring and heterogenization on the one hand, and hierarchization and externalization on the other. It will therefore be governed not only by technical factors, although these are indubitably of the greatest importance, but also by the interests that steer the actors in the network in other ways as well. In other words, technological development is a process combining technical and political aspects and even including a major element of chance.

The technological development of the network provides an important framework for the actions of the individual company. And it is this action which is the main focus of the present study. The questions that we shall be addressing have been formulated in terms of three processes. The first, with which our main research issue is also linked, is concerned with technological development pursued in relationships with external parties. Important issues here concern the extent of this form of technological development and the identity of the most important partners. The second process concerns the development of the base that supports corporate technological development. Here we distinguish particularly between experiential and theory-related technological development. Important issues that we have identified here concern such things as the importance of research-oriented units to companies in general, and the effect on corporate action of the availability and the nature of possible collaborative partners in the immediate environment. The third and last process is concerned with overall corporate development in terms of step-by-step or leap-wise changes. Most technological changes consist of small gradual improvements, but occasionally larger leaps occur. Interesting questions here are: to what extent does external collaboration lead to major leap-wise changes, and in so far as it does, is any special type of partner involved? We shall now try to answer the questions posed in this chapter with the help of an empirical study of inter-company technological co-operation.

Notes

1 In an analysis of a failure of technology transfer during the eighteenth century Lindqvist (1984) concludes that critical factors, in addition to the technical ones, were geographical, economic, social, and cultural. Thus the technology can not be separated and assessed outside the total situation. Lindqvist quotes White (1970: 276): 'My thesis is that technology assessment, if it is not to be dangerously misleading, must be based as much, if not more, on careful discussion of the imponderables in a total situation as upon the measureable elements. Systems analysis must become cultural analysis, and in this historians may be helpful.'
2 The identification of these three processes should be seen as a first attempt to find a suitable structure. Our knowledge is still too limited to make it possible to develop what can be thought of as a final model.
3 Similar views of the development process have been given by Utterback and Abernathy (1975), who relate technical development to the product life cycle; Sahal (1980), who sees it as a learning process; and Steindl (1980), who describes the same process in evolutionary terms.
4 In Håkansson (1987) the process was illustrated with the 'muddling-through process' first identified by Lindblom (1959). The concept 'genuine uncertainty' is used in accordance with Nyström (1974).
5 The heterogeneity of the resources is related to the different effects of combinations of them. This variation is partly due to the complexity of the total situation (see Note 1 above), partly to the uncountable number of possible combinations, and partly to the limitations in our knowledge regarding different dimensions of individual resources. The increase in our knowledge during the last century has taught us a lot about our lack of knowledge – that we still know a fraction of what there is to know about different resources.
6 The use of the resource, and therefore its value and effects, is very much dependent on this hierarchization. This can be seen in many cases regarding developing countries. Galeano's (1976) analysis of the development of South America argues that South America was too rich in terms of resources, which made it an attractive prize. The resources

Industrial networks

were then extracted from South America, giving no development there. In our terms, the resources were not used in order to build up any local networks; instead they were absorbed by the international network with centres in Europe and the United States.

Chapter four

Empirical study

Corporate collaboration on technological questions has been studied in a group of small and medium-sized production companies.[1] In accordance with the theoretical approach presented above, the chosen collaborative mode was regarded as an integral part of the company's overall conduct in the network. In order to study this situation as a whole we needed both general and specific data about development in the relevant companies, about their actions, and about the networks to which they belonged. The task of collecting the data was therefore a lengthy one.

The present chapter describes our approach and some of the problems it involved. The chapter opens with a brief presentation of the plan of the study, and the implications of our chosen design. I then describe the model which determined the design of the interview questionnaire; the theoretical basis of the model has already been discussed in Chapters 1–3. The selection of companies, their willingness to participate, and the final sample are then discussed, and the chapter concludes with a detailed description of the questionnaire used. This questionnaire was our main investigative instrument, and I shall mention below some of the difficulties that arose during the interviews in connection with various parts of it.

General design

Our ambition was to identify and describe corporate behaviour in connection with technical issues and in relation to various kinds of potential collaborative partners such as customers, suppliers, research units, consultants, and so on. We then sought to relate the registered behaviour to other dimensions of corporate action and to examine it in relation to the network to which the company belonged. By describing the different collaborative modes and their relation to other dimensions, we hoped to get some idea of the overall variations; but we also wanted to assess the frequency of the

Empirical study

different modes and to see how far the network situation determined behaviour in this respect. In other words, we wanted to obtain a broad picture of how individual companies handled such questions, and what factors influenced their actions. It thus seemed obvious that our sample should be fairly large – more than 100 companies.

A great deal of the information we wanted was of a qualitative kind, and could therefore only be obtained from interviews with people in top positions. Our main method of collecting data was thus by interview. In addition, we gathered a certain amount of hard data, such as purchasing and sales statistics and the five latest annual reports of the companies. The interviews were well structured, as can be seen from the interview guide in the Appendix. It also had to be possible to feed all the data into a computer, the volume of the data – the large number of companies and the many dimensions studied – excluded the use of any other analytical method. The advantage of this approach – the inclusion of a great many corporate units and the use of well-structured questions – is that it allows for reasonable certainty in postulating the existence or validity of specific phenomena or general connections. The disadvantage is also obvious: the study cannot capture variations in any dimensions not predicted when the questionnaire was designed; the rest disappeared in the standardization process.

The model

The model is a restricted and more specific version of the basic structure for network analysis described in Chapter 2. Its main constituents are shown in Figure 4.1. The focus of the study was corporate collaboration with external units – as a whole and in terms of specific development relationships. This aspect of corporate network behaviour will be related to various dimensions that typify the corporate relationship with the rest of the network. This last is described and analysed mainly in terms of primary relationships or 'organizational sets'.[2] Our chosen method makes it impossible to obtain a more comprehensive description of the network except in one respect, namely the geographical dimension. Companies were selected from a geographically-defined population, which means that we can say something about certain differences between the geographically-defined networks.

In order to describe corporate relations with the network, we distinguish between resources, activities, and actors. The resource structure for the five areas previously identified is described and analysed on a basis of the individual company's role and function in the subnetworks that can be linked to each of the areas. The

Empirical study

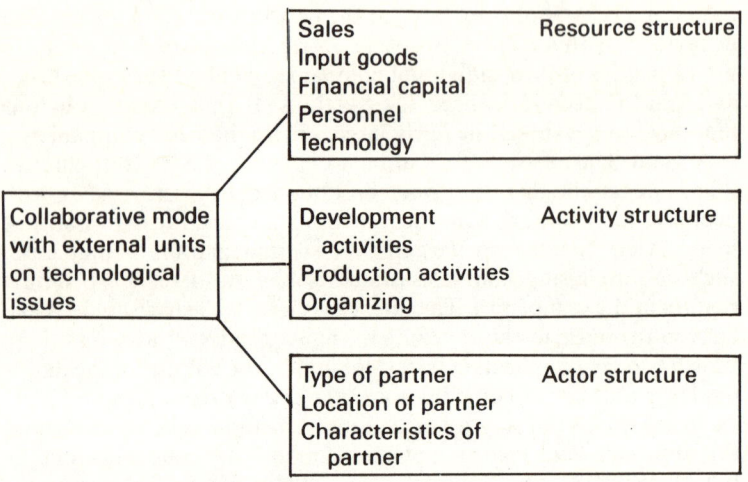

Figure 4.1 Variables studied

company's collaborative mode as a whole and in relation to other counterparts such as customers, suppliers, and research units, is likely to depend in various ways on the nature of the resource areas and the company's role in these areas. The collaborative mode is also likely to be related to activities in terms of total development investment, production, and the design of the organization – all of which represent different aspects of the activity structure. Similarly, we look at the collaborative mode in relation to the type of actor structure involved: for instance, what is the function of the other party *vis-à-vis* the company, where is it located, and so on.

The model in Figure 4.1 has also been used to structure an account of the results. This is started with a survey of corporate network behaviour in terms of collaborative mode by identifying and describing various 'co-operation profiles', and relating these at a general level to influential factors. At the same time the collaborative mode is also linked to two dimensions of corporate results, namely sales and profit.

The three following chapters, in which the results are reported, are devoted to resource structure (Chapter 6), activity structure (Chapter 7), and actor structure (Chapter 8). Chapter 6 is subdivided into sections on the five resource areas and Chapter 7 into three sections on the types of activity included in Figure 4.1. Chapter 8 is a little more comprehensive, since it includes a discussion of the nature and function of the individual development relationships.

Empirical study

Corporate sample and interview subjects

Because of the approach adopted in our empirical study, it was necessary to obtain some idea of each company as a whole and of its role in the network (and various subnetworks); at the same time we wanted to have a large sample. To satisfy both goals without having too unwieldy a study on our hands, we limited ourselves to companies or corporate units of between 20 and 500 employees. 'Corporate unit' can mean whole companies or identifiable parts of some larger company. Further, the study was limited to manufacturing companies. Geographically, the sample was restricted to companies in four counties in central Sweden. In co-operation with the relevant county administration boards, about 130 companies were selected in the counties of Södermanland, Västmanland, Örebro, and Uppsala.

Altogether the selected companies represent a dominating section of production operations, at least in the first three of these counties. The dispersion of the companies over geographical areas, and their size, are shown in Table 4.1. A pilot study was conducted in Södermanland and reported in Henriksson and Håkansson (1985).

Table 4.1 Companies studied

Region	Number of companies approached	Number completed	Number of employees in the completed cases				
			20–49	50–99	100–199	200–299	⩾ 300
Södermanland	40	35	12	6	6	6	5
Örebro	47	47	2	13	14	9	9
Västmanland	30	28	15	5	6	1	1
Uppsala	13	13	6	2	2	2	1
Total	130	123	35	26	28	18	16

In the smaller companies (fewer than 100 employees) one person was usually interviewed, namely the chief executive. In the larger companies several people (but not more than six) were interviewed: the managing director, the marketing director, the purchasing manager, the technical manager, and so on. The interviews lasted from three to four hours in the smaller companies, and up to two days in the larger ones. In addition, the companies provided a certain amount of quantitative data and other supplementary information.

Empirical study

Three of the companies we contacted refused to participate. Altogether about ten companies were unable to take part for a variety of reasons, usually because some special upheaval was taking up a lot of management's time. In seven of the companies interviewed we were unable to get all the complementary material we needed – particularly the necessary economic data. The analysis in the following chapters is therefore based on 123 companies. In some of these cases certain bits of information may still be lacking, but not to such an extent as to disqualify the companies from the analysis.

The questionnaire

The questionnaire (see Appendix 1) was designed largely according to the five resource areas indicated in Figure 2.1. It was possible to integrate the activity and actor dimensions directly under the resource headings. Development and production activities fell naturally under the technical heading, and of the various organizational aspects we included only some of those directly related to external partners, for example distributional features. Similarly, the actor descriptions could be linked to the respective resource areas, especially marketing, input goods, and technology.

The interviews opened with a number of questions about general conditions in the company, so the questionnaire starts in the same way. Questions 1–12, for example, refer to size, turnover, age, location, and ownership. Each question could theoretically be associated with a particular resource area, but generally there were links with several of the areas. These questions were dealt with first because they were easy to answer and on the whole not very important.

The second part of the questionnaire refers to input goods (purchasing). Questions were asked about the purchasing structure for different types of input goods, and about the general appearance of the supplier network (questions 12–61). Questions were also asked here about the suppliers and their importance to technological development, and so on. This section concludes with a detailed review of the three supplier relationships that were most important in the context of technological development, in so far as there were any.

The third section is concerned with sales, and follows essentially the same pattern as the previous section, that is the first part consists of a general survey of sales and the distribution and customer networks, and the second of a detailed description of the three most important customer relationships in the context of technological development.

Section 4 (questions 200–215) is devoted to corporate personnel

Empirical study

structure. Questions were asked about the standard of education, experience, mobility, staff shortages, and levels of competence.

Section 5 is concerned with capital. Questions were asked about ways of measuring capital intensity, ownership structure, capital structure, bank relations, profit, and volume growth.

The sixth and last section is concerned with technological aspects of the company, and questions were asked about production technology, value added, technologial features of the products, and so on. Various important appendix units such as consultants and research institutions were also analysed, and more detailed reviews were made of the three most important relationships in this area. Finally, questions were asked about development intensity, and about the extent of external collaboration in development activities as a whole. The idea was that by the time these questions were asked, the interviewer would already have a reasonably clear and comprehensive picture of the company. The amount of resources invested in 'external collaboration' was particularly difficult to estimate, and this question generally required a good deal of discussion before any answer could be given. Even then it was hard to be very precise, but we considered the reliability of the estimates as a whole to be fully acceptable.

The interviews were conducted by four interviewers over a period of five months. One of the interviewers was responsible for more than half the interviews. All four were familiar with the field, and they met frequently to follow up results, and were thus able to solve any problems of interpretation or interview technique as they went along. Thus, by and large, the data-collecting phase proceeded according to plan.

Reliability of the study

The study was wide-ranging, both in its approach and in the large number of companies that were included in the sample. However, a good deal of the information consisted of facts about the history of the companies in the five resource areas. Obviously, among so much material some may not be of the highest quality, but our general impression throughout was that almost all our respondents did everything they could to see that the information we received was correct. Many companies sent us supplementary information after the interviews, in the form of internal statistics on sales, purchases, and results.

If information failed to materialize or to reach the desired standard, the company in question was excluded from the analysis.

Empirical study

Notes

1 The focus on small and medium-sized companies should be seen in the light of the design of the entire research programme. This also includes case studies of the technical development in larger companies, as can be seen in Håkansson (1987), Laage-Hellman (1988), and Waluszewski (1988).
2 Networks are here defined in accordance with Cook and Emerson (1978) as 'connected exchange relationships'. If these are identified from the viewpoint of an individual company, we get what Evan (1966) has called 'organizational sets'. As an alternative the networks can be identified in relation to a certain technology, resource, or geographical area.

Chapter five

Corporate network behaviour

Network behaviour and corporate identity

The sequence of events in which a company takes part and which it tries to influence occurs within the setting of a network. These events consist primarily of interactions and counter-actions between the companies in the network; but also of the situations and activities generated in individual companies by the various interactions. Each company tries to exploit both the situations as such and the opportunities for different combinations provided by this variety of interdependent situations.

Every situation involves several parties, either directly in exchange transactions, or indirectly in the shape of expected reactions from other parties in future exchanges. The situations are created in part by the company itself, but they owe more to the actions of others. Thus they are not given *a priori*, but are a result of the whole interactive and counteractive process.

The exchanges between the actors provide the crucial base for all change, as well as the point at which the corporate identity is created (Ford *et al.* 1986). At an abstract level every exchange can be regarded as a series of questions and answers such as: 'What can I do for you, and what can you do for me?' 'What do you think about this?' 'I like this, but I don't like that.'[1] The corporate identity is formed and reformed in the course of this continual interaction. It is also in this interaction that the corporate resources are activated and integrated. The exchanges create connections, thus supplying a setting which in turn provides the company with meaning. Thus no individual exchange is an isolated phenomenon, but is always related to other exchanges among the actors in the network. Every exchange is part of a larger whole, and it is this whole that is reflected in various ways in the exchanges and which thus imbues their content with meaning. In other words, the content of these exchanges cannot be fully understood out of context, any more than we can understand a whole play by listening to one line. None the less, our understanding of the play can be affected by the way this or any other line is spoken. This duality, or mutual tension, between the individual exchange and the 'whole' in terms of the network, is one of the most important aspects of network theory. It is also one of the most difficult to analyse.

Predominant in the sequence of events are all the concrete and practical problems connected with inter-company exchanges. Products have to be manufactured, delivered, and used. Each company has to fit into several different activity chains. Individual activities and activity chains have to be streamlined, new combinations have to be created, and old ones discontinued. All these problems are handled within the framework of the exchange processes. Many of the problems – and thus the processes too – are remarkably trivial in themselves, but together they represent by far the greater part of all development activities. It is in this context that companies confront their counterparts with new demands or perhaps with new offers. The individual company may be called upon to make technical or administrative adjustments, to streamline its logistics or its distribution, or to find solutions to a variety of problems. At the same time it may receive diverse offers of products or services, or the chance to join in some common project.

The network is created by the interactions of several units. It is thus the result of the total interplay between these units, and it is constantly being affected by the exchanges and activities undertaken by all the companies involved. To a greater or lesser extent each company takes an active part in the change process by interacting with other units and by undergoing internal changes. We can assume that corporate behaviour is not entirely random but is at least partly controlled in light of the image which the company has formed of the network and of its own identity there. We are thus assuming that an idea has been formed in the company of the nature of the relationship between company and environment. This overall idea, which represents the corporate identity, may be more or less conscious, more or less uniform, more or less historical, and so on, as can be inferred from the way the company behaves. By this I do not mean that all individuals in the company have (or even should have) the same mental picture of the corporate identity. On the contrary, the pictures will certainly vary considerably from one person to another both in content and in their abundance of detail. But taken together, the separate pictures form a certain logical whole that we can call the corporate identity.

The corporate identity defines the company's position and function in the network. The functioning of the network has been described above in terms of activity dependencies and resource linkages. Together these can be said to constitute the basic technological conditions to which the corporate identity is in some way related (or can be related). These basic technological conditions, like the corporate identity, indicate the situation at a particular moment, but they also indicate a direction for change. Technological development thus represents an integral part of both the present situation and future changes. This also suggests that the identity concept refers not only to a mental image, the result of thought, but also to a series of activities which are manifestations of the thought. Perhaps in many cases the manifestations even precede the thought.

In this first section below I shall start by trying to describe the corporate identity of the companies studied as manifested in certain selected dimensions. Second, I shall examine the relationship between these identities and a number of basic corporate attributes. Third, I shall try to relate the chosen dimensions to result variables such as profit and growth.

Propensity to co-operate in technological development

The corporate identity reflects the relationships between a company and various external units. This relationship can be described in various ways, for instance in terms of power and/or specialization. However, my present approach to the description of corporate identity is based on the extent of the integration between company and environment, particularly in the area of technological development. The primary dimension examined here is the company's propensity to co-operate with external units on questions of technological development. Later I shall look at other dimensions such as power and specialization, to explain variations in the propensity to co-operate.

External partners are not, of course, all equally important to the company. In quantitative terms their importance can be expressed as the relative amount of development resources devoted to external collaboration. Table 5.1 shows the relative share of development work conducted in collaboration with external units, according to the companies interviewed. Roughly speaking, about 50 per cent of all development work appears to involve close collaboration with various external partners.

As we can see from Table 5.1, there is a wide dispersion, with only about one-third of the companies in each of the three main categories. We tried to identify the distinguishing characteristics of the companies in the different groups and found, for example, a weak correlation between the level of external collaboration and development intensity. In other words, companies that invest a great

Table 5.1 Relative share of development activities conducted in collaboration with external units

Relative share of development work involving collaboration		Companies	
Low share	0– 9%	7%	
	10–19%	12%	
	20–29%	16%	35%
Medium share	30–39%	8%	
	40–49%	6%	
	50–64%	18%	32%
High share	65–79%	18%	
	≥ 80%	15%	33%
	Total	100%	
Weighted average value: 48.6%			

deal in technological development have a higher proportion of external contacts than companies with a lower level of such investment. In order to explore the relationship with underlying factors, we conducted a step-by-step regression analysis, and tested a number of variables.[2] The most important of these factors affecting corporate development are identified below:

Dependent variables: Relative share of external technological development

Intercept	B value 2.66
Percentage of customer-led production	0.05
Export percentage	–0.11
Stability in customer structure	–0.13
R square	0.144

All independent variables significant at the 5 per cent level.

There are certain general relationships, primarily with attributes of the marketing side but also with the technological nature of the company. The most important independent factor is the relative portion of customer-led production; at the same time this is the technological factor that most explicitly describes the company's general relations with other parties in its environment. It is perfectly natural that the propensity to co-operate on technological issues will be greater in a company whose production is heavily geared to customer orders. Thus the first factor is connected with the company's technological nature in relation to the marketing side. The two other factors that appear to be significant are more directly concerned with the sales side. Companies with a high relative volume of exports were generally less involved in external collaboration, and the same is true of companies enjoying a stable relationship with their ten largest buyers. That a large export share would mean less inclination to collaborate with external units was both expected and unexpected. It was expected in that a company which exports a lot presumably has to standardize its production more than a purely domestic company. But it was unexpected in that the need to make local adjustments tends to produce the opposite effect. However, the final result of our analysis shows that the forces of standardization dominate, manifesting themselves not only in customer relations but also in the company's overall propensity to co-operate.

Stability in the customer structure was measured by determining how many of the present ten largest customers were in the same category five years ago. It was found that greater stability implied less inclination to collaborate with external units. This partly contradicts our basic assumption that development requires stability, and it also conflicts with other results that will be examined below. Let us therefore return to this question later.

We can now add a further dimension to the corporate identity, looking not only at the relative share of external collaboration but also at the direction it takes, in other words, at the corporate co-operation profile.

Co-operation profiles

From the number and direction of the development relationships of the individual companies in our sample, we identified the following four classes of co-operation profile:

- isolated companies (29)
- focused companies (44)
- broad co-operation profile (20)
- very broad co-operation profile (28)

Each of these classes further divides into a number of subgroups, as follows.

Isolated companies   This group has a very limited interface as regards technological collaboration. Of the 29 companies, six have no collaborative relationships at all, five have a few (not more than five) horizontal collaborative relationships, six have similar relationships with customers, and 12 with suppliers. By and large, however, the companies in this group lead a relatively isolated life in terms of technological development activities.

Focused companies The focused companies have a good many collaborative relationships in one direction, but fewer in any other. Of the 44 companies in the group, 28 collaborate mainly with customers, 13 mainly with horizontal collaborative partners, and only three with suppliers.

For various reasons the companies in this group collaborate predominantly with a particular type of partner, perhaps depending on their type/pattern of production. Certain types of subcontractor are included in this group, for example, and they naturally work mainly with their customers.

Companies with a broad co-operation profile Companies with a broad co-operation profile work actively together with a number of partners of at least two of the three main categories identified above. Among these, suppliers and customers dominate. In four of the 20 cases, the companies have extensive contacts with all three categories.

Companies with a very broad co-operation profile This group consists of 28 companies, which all have at least five important collaborative partners in at least two categories. Suppliers are the predominating collaborative partners here, featuring in 26 cases. Customers and horizontal collaborative partners occur in 17 and 15 cases respectively. The companies in this group enjoy close integration with other companies, and represent the extreme opposite of the isolated group.

The co-operation profile of a company defines an important dimension of the corporate identity, and this in turn can be related to the company's situation within the network. To help us to capture at least part of this situation, we have summarized the relationships of the different groups of companies on the purchasing and selling sides (Table 5.2). Taken as a whole, these suggest a clear logical link between the co-operation profiles and the situation in the network. For example, the two extreme groups – isolated and very broad profile – also exhibit the most extreme values in the situational characteristics: nine each, as compared with five and seven for the other two profile groups. In other words, their extreme profiles can be explained at least in part by the situation they occupy in the

Table 5.2 Characteristics of purchasing and marketing in companies with different co-operation profiles

	Co-operation profile			
	Isolated	Focused	Broad profile	Very broad profile
Purchasing				
Share of largest single commodity purchased	**5.6**	*4.1*	4.4	4.5
Share of ten largest suppliers	**7.0**	6.2	6.4	*6.3*
Level of technological development on the market for the products purchased	2.6	**2.7**	*2.4*	**2.7**
Buyer's importance to supplier in terms of volume	**1.4**	*0.7*	1.0	0.8
Import share	2.7	**3.1**	*2.5*	2.7
Share of imports from the Nordic countries	**4.4**	2.3	3.5	*2.0*
Share of new products	*4.9*	5.7	5.6	**6.1**
Marketing				
Share of ten largest buyers	6.2	5.9	**6.8**	*5.8*
Differences among buyers	2.7	2.8	*2.6*	**3.2**
Importance to buyers in terms of volume	**2.0**	1.9	1.9	*1.5*
Importance to buyers in terms of technology	*1.1*	2.1	**3.4**	2.1
Distributors' share of sales	4.5	3.4	**4.7**	*2.3*
Largest competitor's share	**5.1**	3.9	3.5	3.5
Market share	*4.3*	4.7	4.4	**5.0**

bold face = highest value   *italics* = lowest value

network. Companies in the isolated group have the highest degree of concentration on the purchasing side; they are important in terms of volume in relation to both suppliers and customers, but not at all important to buyers, for example, from a technological viewpoint. Furthermore, they have the lowest market share and, relatively speaking, in terms of market share they face the hardest competition. In other words, they are minor actors in well-structured networks. In view of this it is easy to understand how and why the particular co-operation profile has evolved.

The companies with a very broad profile find themselves in more or less the opposite situation. They are market leaders, and their customers are many (low degree of concentration) and varied. They use distributors sparsely. Technological development on the buying markets is rapid, and new products account for a relatively large share of purchases. The companies in this group are thus important actors in more differentiated networks – two factors that together generate both the necessity and the opportunity for collaboration.

The company with a broad profile tends to have a few important customers of a relatively similar kind. These customers also use the company in technological contexts. On the buying side there are no special characteristics. Thus these companies meet a relatively well-structured customer constellation, but because of their own position they none the less acquire a relatively broad co-operation profile.

The focused companies generally seem to have the most fragmented buying pattern: the lowest level of concentration in terms of goods and suppliers, little importance in terms of volume, and geographically-scattered buying (a high import share). On the other hand, the picture on the selling side is close to the average. As we have seen in our earlier description, these companies tend to look towards their customers in terms of technological collaboration. According to the above analysis, this can be largely explained by the possible difficulty in identifying natural collaborative partners on the buying side.

I have used the results in Table 5.2 to depict the average network situation experienced by companies with a particular co-operation profile. Even if all such average analyses suffer from a failure to consider important individual variations, the present analysis does suggest that the co-operation profile is a result not only of the company's own ambitions but also of the activities of other units, which can be related to a great extent to the network situation of the company in question.

The effect of the propensity to co-operation on corporate results

Companies are driven not only by external factors but also by their own ambitions. Two important variables frequently referred to in economic and management literature are profit and volume growth. Although some may question whether these variables are as weighty as neo-classical economic theory, for example, would suggest, there is none the less widespread agreement that they possess a certain importance. We can therefore assume that when companies make their internal analyses regarding development collaboration, arguments based on profit and volume growth will weigh heavily.

The effects of investment in technological development on corporate volume growth and profit make a complicated story. In the short run, for example, investment in technological development has a negative effect on profit. One way of reducing this negative effect is to share the costs with others, that is by collaborating with some external units. However, this collaboration may mean that the company will not be able to exploit fully the long-term benefits of its technological development activities.

The relationship between the relative share of technological development conducted in collaborative development projects on the one hand, and profit and volume growth on the other, is thus extremely complex since it is complicated by three other relationships: that between technological development and the result variables, between the relative share of collaborative development and technological development as a whole, and between this relative share and the result variables.

Profit

The most common problems facing anyone who wants to evaluate the effect of a single action are, first, that there is generally a time lag between action and effect, and second, the result is affected by the way the action in question is combined with other actions. In other words, there are a number of mutual dependencies between different variables (actions) which influence the result. This last problem is often so complicated that we have to disregard it; but at least in the abstract analysis economists generally try to deal with the first problem by distinguishing between the long and the short term. Let us assume that without any external collaboration the company's investment in development is expected to have negative effects in the short run, albeit within reasonable limits, but positive effects in the longer run. What difference might it make if the company launched collaborative efforts with external units?

The complexity is so great that it certainly cannot be easy to identify any simple or general relationships. And yet the relationships are obviously interesting to the individual company, since various options are almost always considered in terms of profit and turnover. Thus what managers think about these relationships is important, since it will affect their behaviour. In our empirical study we did not study management's ideas on the subject, but it seems reasonable to suppose that their ideas can be traced back to the customary cost and revenue curves. Below I shall try to discover which relationships best agree with the established economic analytical approach.[3]

To begin, certain learning costs are incurred. Employees have to learn how to handle collaboration, how to start, develop, and discontinue projects efficiently. To counteract such effects and to ensure that the positive short-run effects will be greater than the costs, a certain volume of external collaboration probably seems desirable. These positive effects consist primarily in a reduction in development costs, partly by sharing them with others and partly by working more efficiently; the units involved will each be able to specialize more, within the structure of the development project.

If a company further increases the external share of its technological development, it will eventually find itself buying more or less all development from outside. The company can then exploit to the full the specialization effect described above. At the same time, however, co-operation costs of various kinds may arise, since the company now has such limited development resources that it can no longer provide a natural base for such activities.

It is important to remember, though, that we have been discussing the relative share of external collaboration. If the company invests more on technological development in absolute terms, it can conduct a much larger amount of external collaboration than a company investing less, without these negative effects being felt.

In view of this we could expect an effect curve in the short run as illustrated in Figure 5.1. Obviously, we cannot decide where on an average the maximum would lie, or how it varies for different technologies, and so on. But we can assume that there will be such a variation.

Long-term effects are much more difficult to evaluate. They also depend even more on the company's success in exploiting the opportunities that collaboration creates in other dimensions. A company engaged in collaborative projects has less influence over the way resulting knowledge or new products or processes are used and commercialized. On the other hand, there are gains: getting closer to and learning more about the other party, and being part of a process

Corporate network behaviour

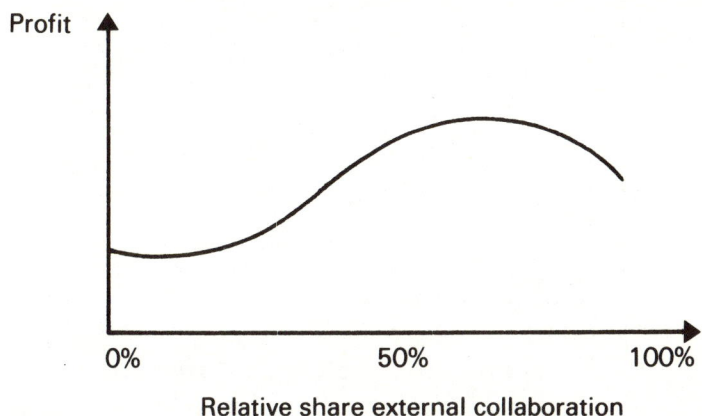

Figure 5.1 Expected short-run effects on profit of different relative shares of external collaboration

that is always opening up new possible combinations. Further, if collaborative development projects do lead to some innovation, more forces are available and can be mobilized to back it up. The outcome of technological development depends not only on the quality of the result but also on how well and how widely it can establish itself; the joint forces behind a collaboratively generated innovation can mean substantial advantages in volume and time. Finally, collaboration can also make for greater flexibility in exploiting the result. Together the companies command a larger total operational arena, with more opportunities for exploiting the final product of the development activities, whatever its form.

An important prerequisite for achieving these positive results is the possibility in the long run of attracting a variety of partners. In order to do this the company probably has to have a substantial internal development activity of its own. When the share of external collaboration increases above a certain level, this will consequently have a negative effect on the attractiveness of the company as a co-operation partner.

For the individual evaluator the final result will depend on the relative weight assigned to the different types of effect. On average, however, the effect curve in Figure 5.2 seems to provide a reasonable picture. Again there is considerable uncertainty as to where the maximum is thought to lie (whether it is at 40 or 60 per cent), but the shape of the curve probably provides a good idea of expectations regarding long-run profit.

61

Corporate network behaviour

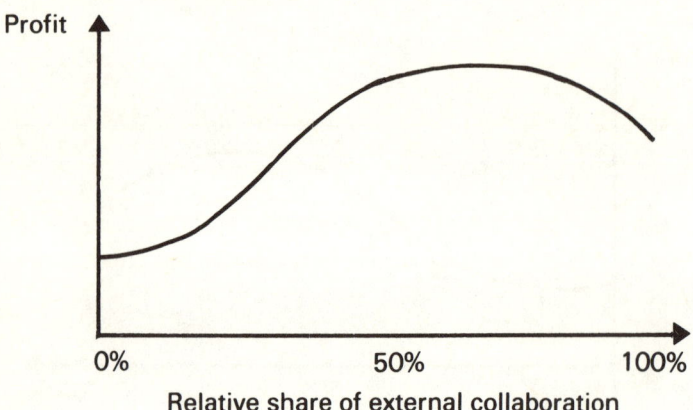

Figure 5.2 Long-run effect on profit of different relative shares of external collaboration

In the empirical study we made some attempt to measure profit in terms of earning power in relation to both paid-up capital and total capital employed. We can thus compare the profit expectations discussed above and identified in Figures 5.1 and 5.2 with actual outcomes in the companies. I shall restrict myself to the relationship between profit for the latest year studied and the relative share of external development activities. By limiting myself in this way, I have a better range of complete interviews with which to work. Profit for a particular year is affected on the one hand (short run) by the relative share of external technological development during the current year, and on the other (long run) by the external share in previous years. In the study the external share had been evaluated for several years previously, so a juxtaposition of external shares and the profit for the last year should tell us something. If the expectations that we have identified are to tally, then profit should be highest in the middle external-share group and lowest in the two extreme groups. Table 5.3 shows that the variation in the data agrees with these expectations, but that the differences are very small.

Table 5.3 Profit at different relative shares of external technological development

External share		Profit – deviation from average[1]
Low:	0–29%	–28%
Medium:	30–64%	+ 4%
High:	65–100%	– 5%

Note:

[1] 18 missing values.

The empirical data provide no grounds for rejecting the type of general expectations that we have identified above. At the same time the results show how difficult it is to establish a direct relationship between profit and the relative external share of technological development.

Volume growth

Theoretically, we could expect much the same picture to emerge in the case of volume growth, except perhaps that the relationship with external technological development should be less complex and more direct. According to our earlier discussion, a higher proportion of external collaboration means better opportunities for mobilizing the network, which in turn should normally favour the volume of sales.

Again we would expect the effect to be most noticeable when internal development investment is still relatively high, and that it should begin to decline when the external share becomes dominating. In the case of volume growth there should not be any threshold effect of the kind we identified in connection with profit. We could thus describe expectations this time as a reversed U-shaped relationship between volume growth and the relative share of external collaboration. Table 5.4 gives the results in the companies studied. The middle group shows the highest growth in volume, and the group with limited external collaboration the lowest. The group with a large relative share of external collaboration occupies a middle position.

Table 5.4 Volume growth at different relative shares of external technological development

External share		Volume growth – deviation from average[1]
Low:	0–29%	–18%
Medium:	30–64%	+ 9%
High:	65–100%	+ 1%

Note:

[1] 11 missing values.

These results accord well with the general expectations that we identified earlier, except for the slight difference that the upper part of the curve does not fall as far as the lower.

Corporate network behaviour

Summary

This chapter opened with a general discussion of corporate network behaviour, and we noted that a company's behaviour is much influenced by its corporate identity. A crucial dimension here is the importance that the company attaches to exploiting the resources of external actors in its own technological development. We have referred to this dimension as the propensity to co-operate, which we first analysed in a general way and in which we found considerable variations among the companies studied. We then examined some determining factors and found that corporate technological attributes and certain aspects of the marketing side were both important. The next identity dimension to be described and analysed was the co-operation profile, which includes both the range and the number of the chosen partners. Four categories of companies, designated as isolated, focused, with broad co-operation profile, and with very broad co-operation profile, were identified and described. The analysis revealed fundamental differences between the situations occupied by the four types of company in their networks. The co-operation profile is thus affected not only by the ambitions of the relevant companies themselves, but also by the way in which external actors behave towards them. Finally, a propensity to co-operate was related to two result variables, namely profit and volume growth. The relationships identified as expected by managers and thus affecting the company's behaviour were confirmed by the empirical data, although the relationship was weak in the case of profit. Some relationships between the propensity to co-operate and the result variables were identified from the viewpoint of managers, that is from their expected perceptions of relationships. The analysis of our data does not give us any ground to reject these expectations.

Notes

1 On a concrete level these questions are interwoven in a stream of events – discussions, deliveries, co-operations, and so on.
2 In total we have tested about 30 main variables characterizing different dimensions of actors, activities, and resources. In the following chapters we will analyse the partial correlations in detail and in several cases they will indicate non-linear relationships. In the overview presented here, however, we limit ourselves to linear relationships.

Chapter six

Resource structure and corporate network behaviour

The handling of resources is an important aspect of network behaviour. All companies (actors) are dependent on resources, and all carve out room for manoeuvres by acquiring control over resources. Technical resources in terms of plant, equipment, and knowledge are a crucial resource area, and one that is obviously related to the company's propensity to co-operate on technological matters. But even the other four resource areas identified in our theoretical discussion in Chapter 2 can have important links with action, since they are closely associated with the technological sphere. We can now analyse each resource area separately, treating them theoretically as five different networks. As the discussion will show, however, it is difficult to draw clear-cut lines between them. The different networks overlap, particularly in the area of technological resources. None the less our classification can help us to obtain an overview of the subject. The resource areas will be discussed in the following order: customers, suppliers, capital, personnel, and technology.

The marketing network

On the sales side the individual company participates in a network of customers, customer's customers, competitors, and in certain cases distributors. The role of distribution will be discussed in the section on organizing activities, and we can therefore concentrate mainly on customers here, and to a certain extent also on competitors.

Let us start with a simple sketch of a possible marketing situation, as shown in Figure 6.1.

At least three questions can be identified from this figure. First, how does the company generally act *vis-à-vis* its customers on technological issues? Second, how are its actions related to customer and buyer structures? For example, should the number of potential buyers affect its actions? Third, how is its behaviour affected by the

Resource structure

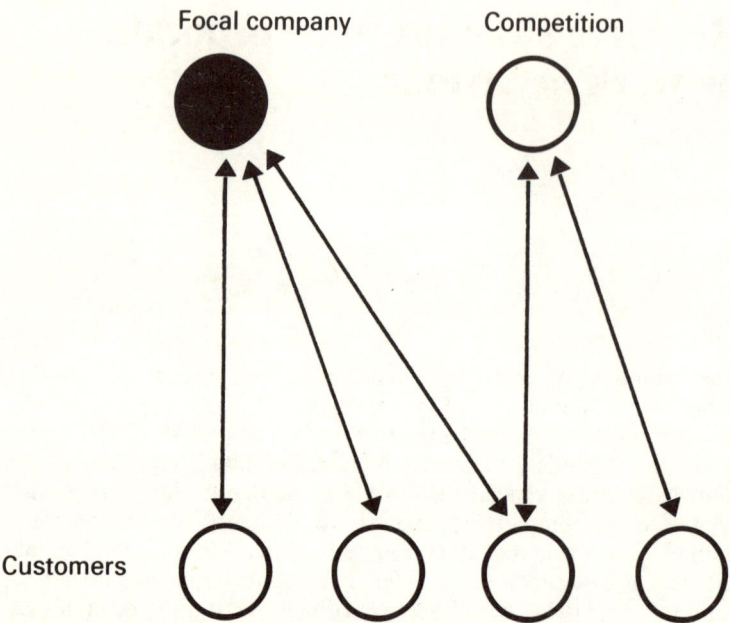

Figure 6.1 The theoretical marketing network

nature of its relations with competitors? Whether or not the company is the market leader, or whether its customers also have relationships with its competitors, are all interesting factors accounting for differences between companies. Each of these issues will be discussed in separate sections below.

Technological attitude vis-à-vis customers

A selling company can behave in different ways towards its customers, and the customers may have different ideas about how they want the seller to behave. For example, does the selling company prefer to work out a unique solution for every customer, and what do its customers feel about this? It is important to ask this question, since the answer affects the company's behaviour.

In various earlier studies (Håkansson and Snehota 1976; Håkansson 1982; Håkansson 1985) we have assumed that a crucial dimension in all customer relations is the amount of individual adaptation that occurs. Thus corporate marketing strategy has been described in terms of adaptive capacity (as regards product or system) and the transmission of this capacity (physical or

Resource structure

knowledge). We have also explored the qualities a company requires if it is to act efficiently at different levels of these adaptive dimensions.

The adaptation dimension was studied empirically both in individual cases, and for a large number of important customer relationships. This has provided us for the first time with a picture of the adaptive dimension and its importance to a large number of companies as a group, which in turn provides us with a test of its usefulness as an initial variable in strategic discourse.

Let us start by studying direct development relationships with customers, and then proceed to examine them in an overall perspective. Figure 6.2 shows how many of the ten most important

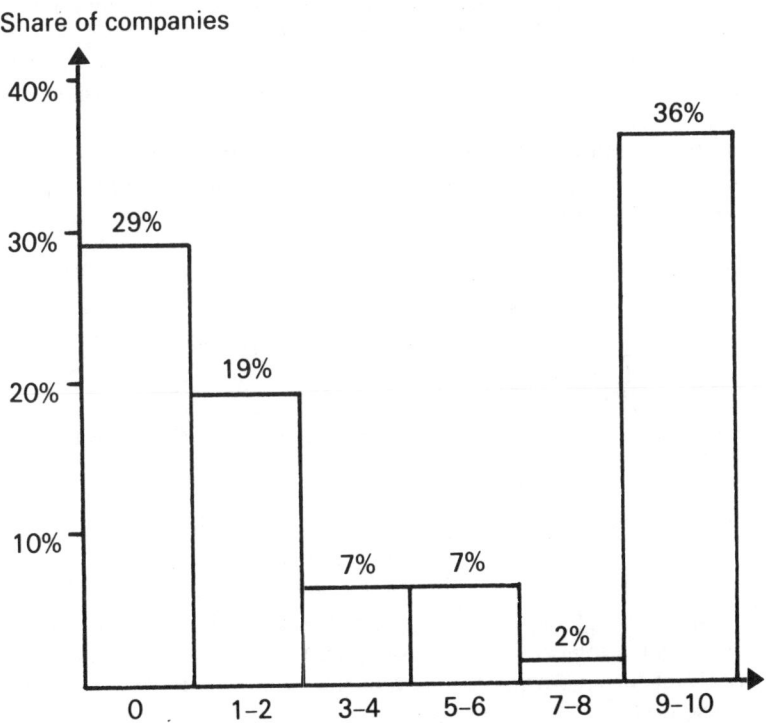

Figure 6.2 Number of co-operation relationships with the ten biggest customers

Resource structure

customer relationships included some form of technological development activity, thus revealing that technological issues are of decisive importance in many major customer relationships in the companies studied.

In over one-third of the companies, all ten major customer relationships included some element of technological development. Less than one-third had no development relationships with customers, and the remainder had a few. Figure 6.2 reveals a clear dichotomy in the group of companies: either companies have many customers for whom technological issues are relevant, or they have few. There is a middle group of about 20 per cent, however, which have development relationships with some but not all of their ten most important customers. In other words, it generally appears to be an either/or question: either a company has technological exchanges with more or less all of its customers, or it has a very small number of forward-directed development relationships. An underlying reason for this is the company's production mode, that is how far production is determined by customer orders. We shall return to this topic in the section on production activities in Chapter 7.

Technological issues in relation to other customers (i.e. those not included among the ten biggest) are also considered important, as Table 6.1 illustrates.

Table 6.1 Technological issues on at least 'test level' in relation to other customers

Frequency of collaboration	Share of companies
Always	8%
Often	8%
Sometimes	19%
Rarely	22%
Never	43%
Total	100%

One-third of the companies sometimes or often conduct fairly close technological collaboration with smaller customers. Over 40 per cent never have such relationships, and a further 22 per cent rarely do so.

Thus technological issues at the collaborative level are important in a great many of the most important customer relationships, but they may also be important in relations with smaller customers. Generally speaking, customers represent an important type of collaborative partner for manufacturing companies; none the less

Resource structure

there is a large group of companies (almost one-third) that does not collaborate with any customers on technological questions.

Development relationships are one specific aspect of a company's technological behaviour towards its customers, but we should also look at customer-related behaviour in an overall perspective. The handling of customers in technological respects varies from one company to another. The seller may not perhaps want to adapt technologically, or the buyer may have no need to demand it. Figure 6.3 shows the variation found in our empirical study. By and large the total group can be divided into three roughly equal subgroups.

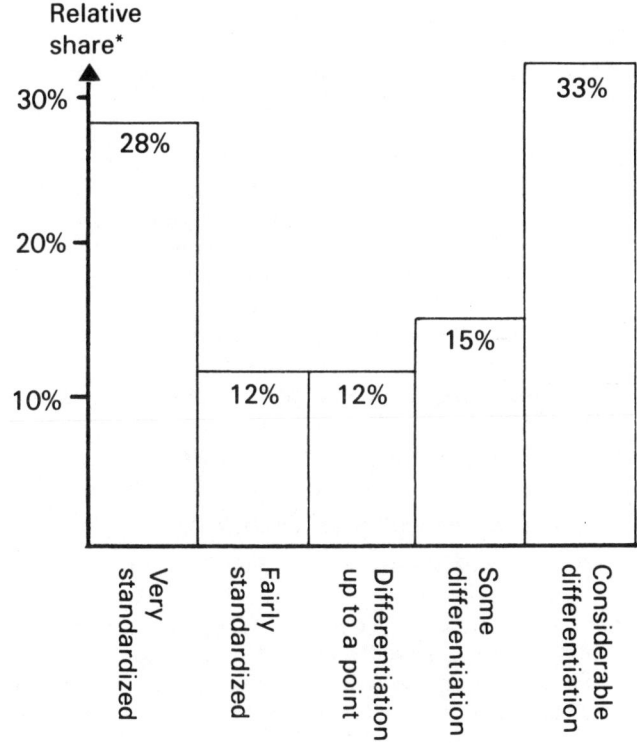

Note:
*Relative share of the studied companies handling their ten biggest customers in the ways specified.

Figure 6.3 Technological adaptation to customers

Resource structure

Almost one-third of the companies exhibit standardized behaviour, treating all their customers the same; one-third differentiates and treats no two the same; and one-third exhibits no regular pattern. The two extreme groups represent a clear and distinct contrast: in one case standardization, and in the other differentiation. The middle group differentiates to some extent among its customers. However, we should be cautious about simply comparing standardization and differentiation, as many companies learn to differentiate in a standardized way. For example, it is possible to create a unique product or system using standardized components or subsystems. So perhaps the difference between the three groups lies primarily in the fact that technical questions are directly involved in the day-to-day commercial exchanges of the two categories noted for their technological adaptativeness. Technology cannot be regarded as given; it represents an action parameter.

We noted above that the present study could be regarded as a test of the usefulness of the adaptation dimension in a strategic perspective. The variation in our sample confirms the importance of this dimension, since the relevant dispersion of the companies proves to be interesting. The two extreme groups each attract about one-third of the companies, while a third group occupies a position in between. We have also noted that the relationship between corporate characteristics and the nature of the customer relationships is simplest and clearest in the extreme cases, while it is obscure for the middle group. Alternatively, we could say that the middle group thus has the greatest opportunities for development. Companies that succeeded in finding efficient ways of combining unique and standardized customer solutions have a broader spectrum to operate in, provided the losses in efficiency are not too great.

Effect of customer structure on the propensity to co-operate

A company's way of working with its customers on technological questions will be affected both by its own nature and by its identity in the network. Since internal characteristics and identity as relating to other resource areas are analysed elsewhere in these pages, we can concentrate here on how the propensity to co-operate is affected by the corporate position on the marketing side.

Let us first examine whether the propensity to co-operate is affected by the number of potential new customers. For instance, we might suppose that collaboration with new customers would be more important to a company enjoying fewer alternatives among its existing customers. However, the opposite would probably apply from the standpoint of the other party: a company with unique qualities as a potential collaborative partner stands a better chance of playing off possible candidates against one another. The results, shown in Table 6.2, do not reveal any strong relationships except in the case of the extreme groups: when the number of potential new customers is very low, the relative share of external development was also low, while the opposite is true when there are a great many potential customers.[1] Thus the number of potential alternatives does not seem to have much effect except in the extreme cases.

Table 6.2 Relationship between number of potential customers and relative share of external technological development

External share		Number of potential new customers					Total
		Very few	Few	Relatively many	Very many	Innumerable	
Low:	0–29%	12	9	14	7	0	42
Medium:	30–64%	9	5	7	11	5	37
High:	65–100%	7	10	11	8	3	39
Total		28	24	32	26	8	118

We could also see whether the total number of customers affects collaborative activity. Table 6.3 reveals an interesting relationship. Where there are fewer than ten customers or more than 100, the frequency of collaboration declines. It is greatest when the customers number between ten and 100.[2]

Table 6.3 Number of customers in relation to the selling company's external orientation

External share		Number of customers			Total
		0–9	10–99	≥ 100	
Low:	0–29%	9	11	23	43
Medium:	30–64%	6	12	20	38
High:	65–100%	2	20	20	42
Total		17	43	63	123

Resource structure

In the middle position it seems to be obvious to both seller and customer that technological collaboration could be advantageous in various ways. The seller will acquire experiences he can exploit in his relations with other customers, and the buyer can gain advantages over competitors.

A third structural dimension that is closely linked to the number of customers, and which could be expected to affect the propensity to co-operate, is the degree of customer concentration, measured in terms of the biggest share of the total sales of a product accounted for by the largest customer and the ten largest customers respectively. The degree of concentration can be expected to affect the propensity to co-operate in two different – and even conflicting – ways. First, greater concentration increases the power of the customers, which makes it easier for them to demand technological collaboration. Second, greater concentration means that the selling company has fewer realistic collaborative alternatives. Both factors can be said to appear in our sample since there is a weak negative correlation between the degree of concentration and number of development relationships on the customer side, but a weak positive correlation with the selling company's total co-operative intensity. A higher degree of concentration means that the selling company has fewer alternatives, but its collaboration with those few customers is all the more intensive.

Effect of competition

A company's situation *vis-à-vis* its competitors can be expected to affect the way it behaves towards its customers. If it occupies a dominating position in its marketing network, it obviously has quite different opportunities for determining its own behaviour – at least as regards the propensity to co-operate – than if it occupies a marginal position (see Table 6.4). More units are interested in co-operating with a leading company, which thus acquires more options among which to choose. Its position is presumably also connected with having access to resources of its own: a leading company will probably have bigger internal resources than one that is more marginal. It has usually achieved its leading position on a basis of certain unique resources, and it is also easier for it to improve its position further once it has become a market leader. At the same time an awareness of this situation often triggers a certain anxiety that others may get at 'our' resources, as well as a feeling that 'since we are the biggest, we must also be the best'. Altogether these reactions may produce a tendency to introversion in that a company with a higher market share will have a lower total external percentage share in its technical development work than companies with more marginal shares.

Table 6.4 The effect of market share on the selling company's propensity to co-operate

External share		Market share			Total
		Small 0–19%	Medium 20–59%	Large 60–100%	
Low:	0–29%	8	16	15	39
Medium:	30–64%	13	9	11	33
High:	65–100%	16	10	11	37
Total		37	35	37	109

The results of the empirical study (Table 6.4) agree with the general picture we expected. On an average companies with low market shares have a greater propensity to co-operate as compared with the market leaders.[3] But although this trend is identifiable, many companies also deviate from the norm. For example, 11 companies with large market shares also have an extremely large relative share of their technological development in external projects. One result that emphasizes the complexity of the situation is the positive (albeit weak) relationship between market share and number of development relationships on the customer side. Maintaining a high relative share of external collaboration can be a way for companies in a weak position to act, just as much as for companies whose positions are strong. It can represent defensive actions in a weak position, or an aggressive way of improving a position that is already strong.

So far we have limited ourselves to one-dimensional situations, that is the relationship between a company and its customers, or between a company and its competitors. In a network perspective, however, we would also expect to find clear interrelationships between these dimensions.

Unfortunately, it is difficult to analyse such complex dependencies in a cross-sectional study of this kind. But we were able to capture one aspect, namely the extent to which a company's ten largest customers also bought from the company's main competitor. Once again we could expect this phenomenon to affect the company's propensity to external collaboration in two ways. First, it may have an inhibiting effect, since the company may fear that what the

Resource structure

customer gains or learns may be passed on to the competitor. Or it may have a positive effect, since the customer has an alternative and more attention must therefore be paid to this company than if the seller was its only supplier.

Our findings on this point are summarized in Table 6.5. The company's propensity to co-operate increases considerably when the most important customers also buy from the company's main competitor.[4] Triangle dramas appear to have a clearly positive effect on the propensity to co-operate. Such triangle dramas can be regarded in turn as a typical product of well-structured networks. In light of these results we would expect this type of network to exhibit a high propensity to co-operate accompanied by a positive effect on development intensity.

Table 6.5 Relationship between the seller's relative share of external technological development and the number of customers (among the ten biggest) that also buy from the main competitor

External share		Number of customers (among the 10 biggest) that also buy from the seller's main competitor			Total
		0–2	3–8	9–10	
Low:	0–29%	17	8	15	40
Medium:	30–64%	5	11	19	35
High:	65–100%	8	9	20	37
Total		30	28	54	112

Results

The technological element in behaviour towards customers varies considerably from one company to another. In relations with the ten biggest customers we found a strong bipolar distribution. Almost one-third of the companies had no development relationships with any of the ten, but over one-third had such relationships with almost all of them (nine or ten of the ten). Obviously, this major difference in customer relations affected the way the company behaved in other respects as well. For example, the companies will face quite different types of strategic problems in the two extreme cases. Where there were no development relationships, it is difficult for the company to keep up with its customers' technological development, and possibly also for it to break through the apparent barriers to collaboration. When there are many development relationships, the situation is the reverse. The large number of relationships may lead to fragmentation, and it becomes important for the company to find means of integration. In the long run it may also be necessary to become highly selective in choosing customers.

The customer structure, in terms of both existing and potential customers, has a certain effect on the propensity to co-operate. The number of existing customers has the stronger effect: when there are between 10 and 100 the propensity to co-operate is clearly greater than when customers number fewer than 10 or more than 100. The number of potential customers affects only the extreme cases, 'very few' customers and 'innumerable' customers. If the number is very small, the propensity to co-operate is affected negatively; if it is very great, the effect is positive. All the states between these two extremes appear to be largely indifferent with regard to the propensity to co-operate.

Competition affects the propensity to co-operate. If a company is a market leader, its total propensity to co-operate declines. However, this tendency occurs primarily not on the customer side but in relation to other external parties. Finally, in a survey of customer contacts with competitors, we found to our surprise that as the number of customer contacts with the seller's competitors grew, the seller's propensity to co-operate also increased.

Thus, to summarize our results, we found that the effect of the customer and competitor structure on the propensity to co-operate appears to be greatest in typical network situations, that is when there is a fairly (but not too) limited number of actors involved in many relationships with one another.

The input network

Suppliers can vary in importance for many reasons. First, it is a question of how much of its production process a company wants to do itself. One company may want to be responsible for the whole process, while another leaves production largely to its suppliers. The latter option is naturally only possible if suitable suppliers are available, and it is just this – the existence of suitable suppliers – that is the second factor influencing the choice between buying and production within the company. A third influential factor is the company's position in the processing chain. Suppliers will presumably be more important if the company operates at a later stage in the process. If, on the other hand, the company operates near the beginning of the chain, the suppliers may be important in volume terms, since companies at that stage are generally raw material-intensive, but suppliers are probably not as important in the context of technological development.

Resource structure

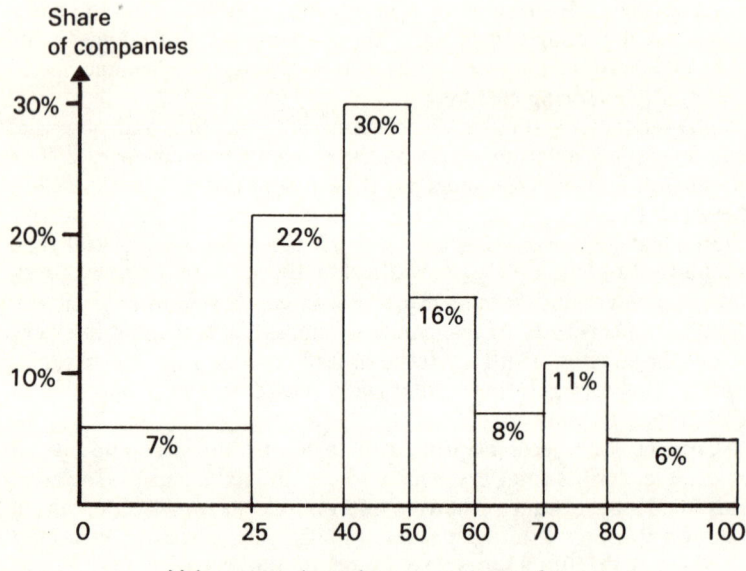

Figure 6.4 Volume purchased in relation to turnover

The group of companies that we are describing were all manufacturing companies, but the volumes purchased varied none the less, as can be seen in Figure 6.4. To a quarter of the companies, suppliers were very important in terms of volume, and these companies had a buying volume exceeding 60 per cent of total turnover.

For almost 30 per cent of the companies, suppliers account for less than 40 per cent, that is they are not very important. Just under 50 per cent have a buying volume varying from 40 to 60 per cent of turnover. Thus for a clear majority of the companies, suppliers play an important role in volume terms.

These suppliers are handled by the buying function of the company, which may be – but is not necessarily – a buying department. The role of the buying function in the company has been described elsewhere in terms of its development role, its rationalizing role, and its structural role (Axelsson and Håkansson 1984; 1986). The term 'development role' refers to the responsibility of the buying function for the exploitation in various ways of the development capacity of the company's suppliers, either by buying products that incorporate new technology or by entering into various forms of technological collaboration.

The rationalizing role includes all efforts to improve the efficiency of the relation between the suppliers and the buying company, perhaps by finding the best balance between what the suppliers do and what the company does. This role also includes efforts to find more efficient suppliers. Thus rationalization may mean finding better suppliers or improving the relationship with existing suppliers, perhaps in a variety of collaborative projects.

Finally, the structuring role represents a monitoring process whereby the buying function reviews the company's evolving supplier network, perhaps making sure that it consists of an appropriate company mix. Collaborative projects with individual suppliers, for instance, might be one way of supporting a company that has fallen behind, but which could still be worth maintaining in the network to ensure alternative sources of supply in future.

Thus technological collaboration does not belong exclusively to the development role; it can equally well appear as part of the other two roles. The number of projects thus tells us something of the overall intensity of buying activities.

The following analysis of technological collaboration with suppliers starts by looking at the number of suppliers with which the companies collaborate, and proceeds to examine the effect of the supplier structure on the form that the collaboration assumes.

Development relations with suppliers

As we have seen, the purpose of development relationships with suppliers can vary, but the existence of such relationships shows that a company is actively trying to exploit the input element in its overall network behaviour. Figure 6.5 shows the number of development relations with suppliers in the companies studied. A great majority have fairly close technological co-operation with 1-5 suppliers.

A relatively large group – a quarter of the total – does not collaborate with any suppliers. A very much smaller group – a sixth – collaborate with more than five of their suppliers. There are two notable differences here as compared with the customer side. First, the dispersion is now more evenly bell-shaped. There is no sign of the dichotomy that appeared in the customer diagram. The total differences are thus smaller among the suppliers. Second, on average the number of collaboration partners is somewhat lower: 3.2 as against 4.5 for the customers. This difference is surprising, and requires further examination. Since the companies studied are all both buyers and sellers, and since buyers and sellers are needed in equal quantities to create development relationships, statistically speaking there should be as many sellers as buyers involved in development relationships.

Resource structure

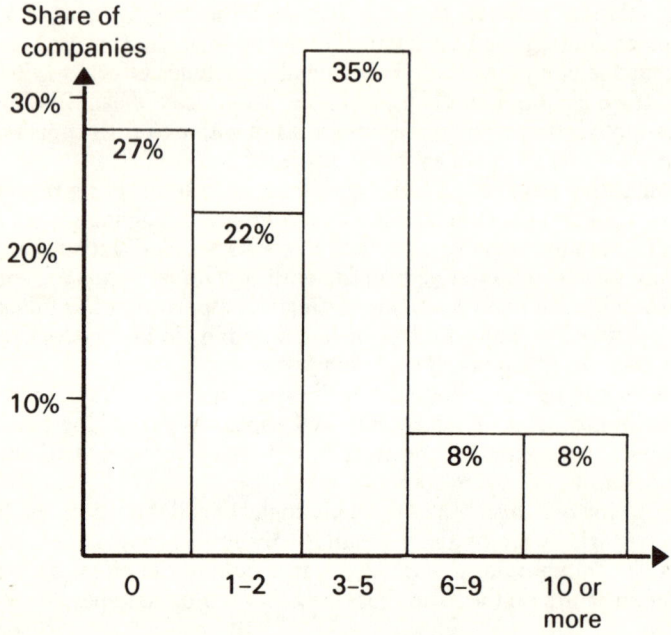

Number of important co-operation relationships with suppliers

Figure 6.5 Number of co-operation relationships with suppliers

One explanation could be that the supplier side is more concentrated. Let us therefore look at Figure 6.6, in which the degree of concentration on the supplier and customer sides is compared. It appears from this comparison that both the customer and supplier sides are normally highly concentrated.

In about 50 per cent of the companies, the ten largest customers account for over 80 per cent of turnover, and for an equally large group the same applies to the dispersion on the buying side. A weighted average for the ten largest customers comes to 70 per cent on the customer side and 72 per cent on the supplier side, that is the difference is marginal. The same applies to the single largest partner, which in both cases accounts on average for about a quarter of the volume. In other words, the degree of concentration cannot explain the difference.

Another possible explanation could be that the size dispersion of the companies studied may have affected the result. The sample consisted largely of small and medium-sized companies whose customers are often larger companies, that is the companies in our

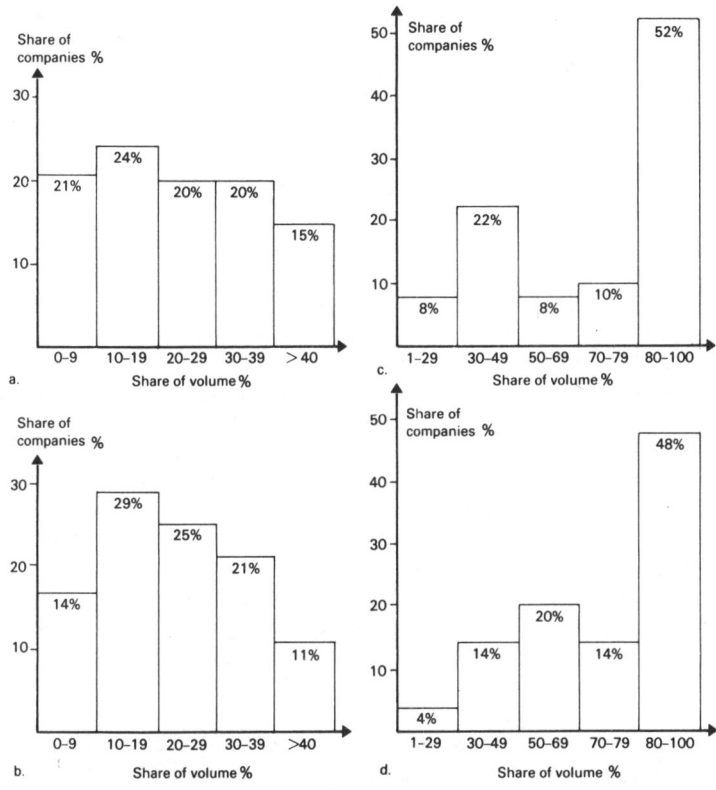

Notes:
a. the largest customer's share of total volume sold (average share: 26%)
b. the largest supplier's share of total volume bought (average share: 26%)
c. the ten largest customers' share of total volume sold (average share: 70%)
d. the ten largest suppliers' share of total volume bought (average share: 72%)

Figure 6.6 Degree of concentration on the customer and the supplier sides

sample operated as subsuppliers. Such companies are more or less compelled to co-operate on technological issues when the customer demands it. As buyers their own size is an obstacle; should they try to interest other suppliers in a joint development project, their volume is often simply not big enough.

A third explanation may be connected with the processing chain. In our introductory discussion we noted that technological collaboration with suppliers was more important to companies at the later stages in the processing chain. By definition, a company's sales function is always further along the chain than its input function. In

Resource structure

view of this, the difference in the number of collaborative relationships on the customer and supplier sides seems quite natural.

Finally, the difference could perhaps be due to the fact that the seller is particularly active and that the selling function may therefore perceive the relationship as a development collaboration (which it is to them), while the buying function does not see it in the same way. Another reason could be that less is known about such activities on the buying side, apart from the buying function itself; these activities are so much part of the day-to-day work that it rarely occurs to anyone to discuss them in these terms. On the selling side such collaborative activities have an impact on other parts of the company, and they therefore become known. The perceptions of the companies are also more attuned towards selling, which may mean that projects connected with the selling function attract more attention.

Thus, to summarize our results, we found that suppliers represent an important group of external partners, although on average the companies have slightly fewer development relationships with suppliers than with customers. The concentration ratios on the two sides are similar so this cannot explain the difference. The explanation is more likely to stem from the nature of the corporate sample, or from the fact that on average the customer side is always more technologically intensive, or that collaboration with customers is more visible than collaboration with suppliers. Looking at the number of development relations, and taking into account the possible aims we have noted, we can conclude that there should be considerable potential in almost all companies for exploiting untapped opportunities for collaboration on the supplier side.

Different types of supplier

Suppliers can be divided into subgroups according to the type of goods they supply: materials, components, and equipment. Technological development can consist of companies buying new products from these three groups. Table 6.6 shows new products as a percentage of total purchases over a five-year period. Not unexpectedly, it is mainly on the equipment side that the bigger changes appear. It should be noted that Table 6.6 refers to radically new products only, so that minor ongoing changes are not included. This last type of change is more common in materials and components which are being bought continually. Equipment, which is bought at fairly long and irregular intervals, generally has time to change more from one purchase to the next.

The 44 per cent of the companies that are registered as buying no

Table 6.6 New products over a five-year period as a percentage of total purchases: materials, components, and equipment

	Share of companies buying		
Percentage of total annual purchases	New raw or processed materials	New components	New equipment
0	64%*	66%	44%
1-4	18%	20%	26%
5-9	13%	5%	10%
≥10	5%	8%	20%

Note:
*For example, 64 per cent of the companies have 0 per cent new raw materials, and so on.

new equipment may not have made any major investments over the last few years. However, if we analyse the type of supplier that is most often chosen as a collaborative partner, we find – rather surprisingly – that these are the suppliers of raw or processed materials. These account for almost 50 per cent of all identified supplier collaborations. The suppliers of machinery and equipment account for barely 30 per cent, and the suppliers of components for a little over 20 per cent. If we compare this with the results in Table 6.6, a reasonable conclusion seems to be that most of the collaborative development projects are concerned with continual adjustments to existing products. This result confirms the picture we previously found in other projects. Development relationships are important in the case of materials such as steel (Håkansson 1979), but also in other cases as well (Håkansson 1982; Turnbull and Valla 1985).

The supplier structure and the propensity to co-operate

The supplier structure can be expected to affect a company's propensity to co-operate in much the same way as the customer structure, although the influence might perhaps be less strong, since most companies are geared particularly towards marketing and this automatically reduces the influence of the suppliers.

We can now examine two aspects of the supplier structure. First, what effect will concentration on the input side have on the propensity to co-operate? Second, how will this propensity be affected by the number of alternatives that are available to the most important supplier? As on the marketing side, the degree of concentration can be expected to affect the propensity to co-operate in two ways. One of these is positive, in that greater concentration throws the spotlight on possible collaborative partners and thus makes the need for collaboration more obvious. But the other is negative, in that the number of potential partners is reduced, and the company's position *vis-à-vis* the suppliers is weaker.

The empirical material indicates a very weak negative relationship between the propensity to co-operate and the degree of concentration. However, the correlation is so weak that it cannot be called a trend and thus there is no point in discussing it further.

With regard to the number of possible alternatives to the most important suppliers, the situation is similar. This factor does not seem to affect the propensity to co-operate to any great extent. Thus development collaboration on the supplier side seems to be relatively unaffected by the nature of the supplier structure. This may be because most companies are still largely unaware of the need to exploit their suppliers efficiently, and it is thus a matter of chance or individual initiative that decides things rather than any structural characteristics of the network. Furthermore, it remains to examine the way in which the external characteristics of a company can affect the external orientation of its technological development activities.

Capital structure

A company's capital structure reflects its relations with owners and lenders. With the help of the simplified network structure in Figure 6.7 we can formulate several questions, starting with one about owner or owners. What kind of owners are there, and how many? If the owner is a manufacturing business, for example, the relationship between company and owner is obviously likely to have a technical content, and we see the type of indirect dependency relationship which owner A has with company X in the Figure. The more owners a company has, the more likely it is that such indirect dependencies will occur with accompanying technological collaborations.

Another question concerns the situation between owners and lenders. The role of the banks in industrial networks has been discussed elsewhere (e.g. Hermansson 1982 for Swedish conditions). If a particular bank has a considerable interest in a company's development, it may also become actively involved in various ways in the company's technological development, perhaps by activating a relationship with some other party, such as company Y in Figure 6.7.

We can now try to discover whether the ownership structure noticeably affects the relative share of external technological development. First, in Table 6.7 we relate the ownership situation to

Resource structure

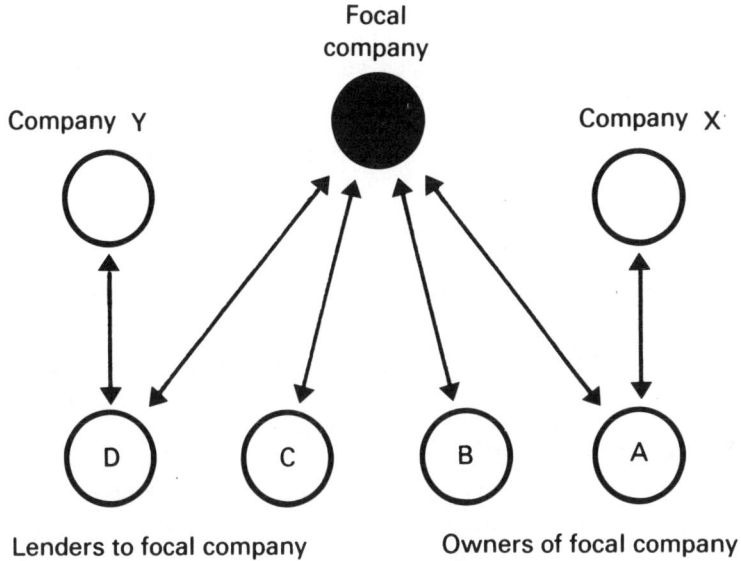

Figure 6.7 The company and the capital network

Table 6.7 Ownership and the relative share of external technological development

Relative external share		Wholly independent unit		Independent division or subsidiary in a corporate group		Production unit in a corporate group		Total	
Low:	0–29%	20	(34%)	13	(28%)	8	(61%)	41	(35%)
Medium:	30–64%	15	(25%)	19	(42%)	4	(30%)	38	(32%)
High:	65–100%	23	(39%)	13	(28%)	1	(7%)	37	(31%)
Total		58	(100%)	45	(100%)	13	(100%)	116*	(100%)

Note:
*Seven missing values.

the propensity to co-operate on technological questions.

As we can see from the table there is a clear difference in the propensity to co-operate as between production units and independent companies or units.[5] As a whole, the production units seem to be much more isolated than independent units, since we even included collaboration with a parent company as part of the external share. On the other hand, the presence of a group as owner does not

83

Resource structure

appear to have any impact, either positive or negative, on the volume of external collaboration on technological questions, compared with the situation for a wholly independent unit. But, as Table 6.8 shows, the number of owners does affect the size of the external share.[6]

Table 6.8 Relative share of external technological development and number of owners

External share	Number of owners						Total	
	1		2		>2			
Low: 0–29%	21	(38%)	10	(33%)	7	(28%)	38	(34%)
Medium: 30–64%	21	(38%)	5	(17%)	9	(36%)	35	(32%)
High: 65–100%	14	(24%)	15	(50%)	9	(36%)	38	(34%)
Total	56	(100%)	30	(100%)	25	(100%)	111	(100%)

Where there is more than one owner there also generally seems to be more emphasis on external technological development. This result confirms our earlier conclusions.

The relationship with lenders also tends to affect this factor. The companies that have the lowest proportion of paid-up capital (less than 10 per cent), and thus the highest proportion of borrowed capital, are the most introverted. The most extroverted are the companies whose paid-up capital is in the range of 10–20 per cent, while companies with a higher proportion of paid-up capital occupy a middle position. This U-shaped relationship could be explained as follows: if paid-up capital is excessively low, the company is presumably in some difficulties and is being kept too busy with its internal problems to have time for external development projects. Otherwise the same relationship would apply as indicated above, that is a higher proportion of borrowed capital would be accompanied by greater pressure to take part in various kinds of external projects.

Personnel structure

Personnel can be regarded mainly as an internal resource. However, each individual staff member also has a personal network of contacts and thus represents a crucial extroverted dimension in corporate activities.[1] All employees have a life of their own outside the organization. They generally only spend about 20–25% of their time as an employee; the rest is spent together with family or friends. Even if a good deal of time is taken up by sleep, a lot is still left for other private activities. This private activity may be relevant to the

Resource structure

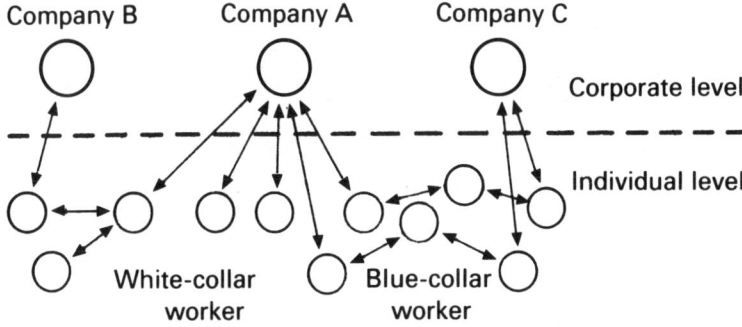

Figure 6.8 Personal networks

company to a greater or lesser extent. Private contacts can be of greater or lesser value to it. For example, a qualified engineer may keep in touch with numerous fellow students who get posts in firms which may be of interest to his own company. Figure 6.8 illustrates this kind of situation.

Obviously, it was not possible in the empirical study to obtain information about the importance of individual people in this respect, but we recorded the number of staff, the distribution between blue- and white-collar workers, their standard of education, experience, and mobility. All these dimensions can be related to the company's propensity to co-operate on technological issues.

Total number employed and service structure

The relationship between the total number employed and the relative share of external technological development is certainly a very complex one. Generally speaking, we could expect that as the number of employees rises, it will be more difficult to engage every one of them in the establishment of external links. On the other hand, the number of employees is also a measure of resource strength, and as such should encourage external parties to seek collaboration with the company concerned. Table 6.9 shows the results of the empirical study on this point: the results are interesting and provide a good illustration of the complexity concealed within variables such as 'number employed'.

The smallest companies exhibit a very clear bipolar dispersion: the companies are either very outward-looking or obviously introverted. Few of them belong to the middle group. Among the companies with the most employees (over 200), on the other hand, the middle group is the largest, although there are also many introverted companies

Resource structure

here as well.[8] The highest average extroverted character is to be found in the group with 100–200 employees.

Table 6.9 Relationship between number employed and relative share of external technological development

External share		Number employed				Total
		20–49	50–99	100–199	≥ 200	
Low	0–29%	13	10	6	13	42
Medium:	30–64%	4	7	11	16	38
High:	65–100%	17	9	10	3	39
Total		34	26	27	32	119*

Note:
*Four missing values.

It is clearly difficult for a company with a great many employees to maintain a high external share. The very small companies adopt either of the two extreme positions; it is possible for them to choose, and they appear to do so. The middle group finds it more difficult to steer events and thus also to choose, and they are certainly influenced by their technology and their network.

The second dimension refers to the division between blue-collar and white-collar workers. This dimension, too, reflects several underlying conditions. One is that white-collar workers are generally more involved in questions connected with technological development, and their contact network is more interesting in this respect than that of the blue-collar workers. In view of this we might expect that the greater the proportion of white-collar workers, the greater also the relative share of external technological development. At the same time, the proportion of white-collar workers also reflects the technological structure of the company. Their proportion rises when production is streamlined and increasingly automated, or when the service element in the final product is increased. In the first case production often becomes less flexible, and in turn puts great pressure on relations with suppliers and customers; this therefore seems likely to inspire a bigger relative share of external technological development. In the second case the company often becomes more flexible, which in itself implies more opportunity for collaboration per se, but also means that there is less demand for it. These relationships are illustrated in Table 6.10.

There appears to be a U-shaped relationship here.[9] Companies with a very low proportion of white-collar workers, under 10 per cent, generally have a very low share of external technologial development. The share in companies with a high proportion of

Table 6.10 Relationship between relative share of external technological development and proportion of white-collar workers

External share		Proportion of white-collar workers				Total
		0–9%	10–19%	20–29%	≥30%	
Low:	0–29%	5	12	14	11	42
Medium:	30–64%	2	15	12	9	38
High:	65–100%	1	13	18	7	39
Total		8	40	44	27	119*

Note:
*Four missing values.

white-collar workers is almost as low. The middle groups have the highest rate of external technological development. These results therefore suggest that a high proportion of white-collar workers does not guarantee a strong external element in technological development and that the basic technological conditions are more important in this respect.

Education

Education is often intimately linked with technological development. We looked at the level of education in the companies studied to see whether there is any direct connection with the frequency of collaborative projects. Our description and analysis included the employees' basic education in both general and technological terms, and the intensity of internal training. In many of the companies the proportion of staff with a high-school education (12 years of schooling) or more was low. In more than half the companies the figure was under 15 per cent. The level of theoretical technological knowledge was also low. In almost half the companies (43 per cent) the number of engineers who had completed a course at technical college or at a higher level was less than 5 per cent. The relationship between this situation and the intensity of collaboration is not very clear. Generally speaking, we could expect that a higher level of general education would be associated with the exploitation of external units, but it has not in fact been possible to identify any such simple or direct correlation. In relation to technically trained personnel, there is even a negative correlation. On the other hand, there is a clearly positive correlation between total development intensity on the one hand, and general and technical education on the other. Thus, on average, higher education does not seem to trigger a generally greater receptiveness to technological collaboration. Turning to internal training activities, however, we find that a different picture emerges.

Resource structure

As can be seen in Table 6.11, the number of development relationships rises as the level of activity in internal training increases. This training activity is obviously associated more directly with the company's collaborative mode than with the basic education of the employees. This is not perhaps surprising, since both internal training and collaborative mode are connected with the total activity level in technological development.

Table 6.11 Relationship between internal training activity and collaborative relationships

Share of employees receiving internal training during the past year	Percentage of companies	Average number of development relationships (index)
0–9%	39%	96
10–19%	28%	94
20–39%	17%	107
≥ 40%	16%	113

The average level of internal training is not particularly high. In two-thirds of all the companies included, a maximum of one employee out of ten had taken part in any internal training activity during the past year.

Experience

Experience is an interesting variable in a network context. A network embraces a series of complex interdependencies, of which many are of an indirect kind. Every network is thus unique, and the only way to handle it is to work inside it. Thus greater experience should mean more capability for handling technological collaboration as part of network behaviour. This in turn should mean more interest in collaboration and thus also a higher frequency of collaborative projects. Generally speaking, there should be a positive correlation between the propensity to co-operate and experience. At the same time, the experience dimension is not easy to measure since experience does not automatically increase with time, although it is strongly time-dependent. A systematic analysis of earlier events is needed if experience is to continue to accumulate; if an event is to yield something extra in the way of experience, a certain willingness to experiment is also required. In other words, two companies whose personnel structure is the same in regard to the number of years worked, may be operating at quite different levels in terms of total experience. In the empirical study we had no possibility of allowing for this. Only time-related or assessment-based measurements of the experience variable were possible.

Employee experience was first registered in terms of average length of employment (see Table 6.12), and then assessed by the companies themselves. Respondents were asked whether the level of competence in various functions had changed over the last five years, i.e. whether experience and knowledge were more important now than they had been five years before.

Table 6.12 Average length of employment of blue-collar workers

Number of years	Percentage of companies	Index for relative share of collaborative relationships
0–9	34%	99
10–14	39%	108
≥ 15	27%	89

In two-thirds of the companies the blue-collar workers have been with the company for an average of more than ten years; the same also applies to the white-collar workers. If there is any correlation with the propensity to external collaboration, it is a very complex one. The figures in Table 6.12 show that most collaborative relationships are to be found in the middle group. Thus, up to a certain level, experience has a positive effect; above this level the effect tends to be negative. This may well be because the increase in experience does not operate over such long periods, or because experience then becomes an obstacle rather than a positive asset.

The level of competence was generally judged to have increased somewhat, or noticeably, in the four corporate functions observed. In particular the competence level in the development function was said to have risen. In production and sales it had also increased, but not to the same extent. The competence level was judged to have risen least in the buying function (see Table 6.13). Thus competence requirements were felt to have increased primarily in the technological functions, which had become either more important or more difficult to handle.

These assessments of the level of competence are interesting since they reflect the company's view of what it finds important at the moment, and the way in which it would like to handle related development issues. The answers suggest that it is of primary importance to the company to develop and control technology.

Resource structure

Table 6.13 Perceived changes in competence level in four corporate functions

Corporate function	Index*
Production	3.3%
Purchasing	2.8%
Sales	3.2%
Development	3.5%

Note:
*The values have been obtained by combining the answers, where 1 = declined, 2 = stable, 3 = increased somewhat, 4 = increased noticeably, 5 = increased substantially.

Further, this should be done within the technological field itself, that is by recruiting more highly trained technicians. However, the companies do not seem to be aware of the opportunities for technological development that could be provided by relationships with their customers and suppliers, if the corresponding functions (sales and purchasing) were further developed. Or perhaps management believes the opportunities are already being fully exploited, or that the technical functions should handle such questions as well.

These speculations indicate something of the arbitrariness that we perceived in the way our companies viewed the subject of technological development – its importance and its complexity, and ways in which the problems could be solved. I shall elaborate my own view of this situation in Chapters 10 and 11.

In conclusion, the experience dimension appears to affect a company's propensity to co-operate, and it appears to be increasing in importance with regard to technological questions.

Personnel changes

An important activity in the building up of knowledge is recruitment, for example of engineers or sales personnel. The need for deeper knowledge, and thus the recruitment requirements, will vary for different types of technological development. If development is primarily research-based, there will be more need for researchers who have experience and contacts either from customer companies or from research institutes and university departments. Such organizations provide not only research experience, but normally also extensive contact networks through their many former students who are now out in working life and scattered throughout the network. If, however, a company's technological development is largely experience-based, it might be better to recruit engineers or buyers who have previously worked with the users of the product or process that their new company sells. If a particular input, perhaps a certain material, constitutes a technically important part of the product, then engineers or salespeople from supplier companies would make useful recruits.

In the empirical study we explored two aspects of personnel change. The first was the total level of recruitment and loss of personnel, and here companies were asked whether staff mobility had increased or diminished over the year. The second concerned the difficulty of acquiring the right sort of personnel, and companies were asked to say whether they suffered from a shortage of engineers or skilled workers. The results can be seen in Tables 6.14 and 6.15.

Table 6.14 Staff mobility (percentage of companies)

Changes in staff mobility over previous year	Blue-collar workers	White-collar workers
Increased substantially	0%	1%
Increased somewhat	6%	8%
Unchanged	49%	65%
Decreased somewhat	27%	20%
Decreased substantially	18%	6%
Total	100%	100%

Table 6.15 Shortage of engineers and skilled workers (percentage of companies)

Perceived supply	Engineers	Skilled workers
Very great shortage	5%	11%
Great shortage	13%	18%
Some shortage	17%	7%
Slight shortage	19%	19%
No shortage	46%	45%
Total	100%	100%

Generally speaking, the companies felt that mobility either had not changed or had decreased. Only a few felt it had increased. Mobility had decreased particularly among the blue-collar workers.

Over half the companies felt there was a shortage of engineers and/or skilled workers. The shortage of skilled workers was perceived as greater, which is also suggested indirectly by the figures for mobility.

Resource structure

If mobility continues to be low, more investment in internal training will be needed in the future, and altogether a more long-term programme for building up internal competence. We have already noted (Table 6.11) the relatively low level of internal training activities. In other words, there does not seem to be any great awareness of the problems that we identified in our company sample, or at least there is not enough energy in tackling them.

The technological network

The technological dimension is central to the theme of the present book, and the relative share of external technological development represents a crucial aspect of behaviour in relation to the network as a whole. We have already explored the way in which customers and suppliers are exploited as partners in a technological context, but there is also a third type of partner that we have described as 'horizontal'. In the present section I shall discuss this type of partner, and try to discover how two special dimensions of corporate technology affect the propensity to co-operate. Both are general corporate characteristics, but at the same time they give some indication of the network in which the company operates. The first dimension refers to the role of technology in a production perspective and the second to its importance in the products sold.

Horizontal units

A particular company can collaborate with all kinds of companies or organizations which could be described as 'horizontal'. For instance, there are organizations, such as university research departments and other research institutes, that specialize wholly in producing and communicating new knowledge. There are also national and local organizations such as the National Board for Technological Development in Sweden, or various development funds, and ministers of trade and industry, and so on, whose tasks include helping and supporting technological development in companies. Then there are companies which manufacture competing or complementary products. Finally, there are companies that could almost be described as suppliers, such as the various types of consultancy organizations.

Among all these various types are many which we could conceive of as suitable collaborative partners in a technological development context. These would include, for example, all the units created by society to contribute to corporate technological development, as well as the units that could be said to symbolize the development of

Resource structure

knowledge, namely universities and other research institutes. None the less, the companies in our study had fewer collaborative partners in this horizontal category than they had among their customers and suppliers. Figure 6.9 shows that the biggest group of companies was the one that had only a limited number of horizontal development relationships, typically between one and three (the average was 2.4). Only 15 per cent of the companies had more than six development relationships with horizontal partners. Turning to the different types of horizontal partner, we have no dispersion for all collaborative relationships, but for the most important (which was a maximum of three for each company) we can say that about 10 per cent of them were with university departments of some kind. These relationships are on average not more important than others, which allows the conclusion that universities play a minor role as partners for these companies.

The horizontal units represent a motley crew, and it is therefore difficult to identify any specific horizontal network structure. In the remainder of this chapter we will therefore look at the way two general technological dimensions affect the overall propensity to co-operate.

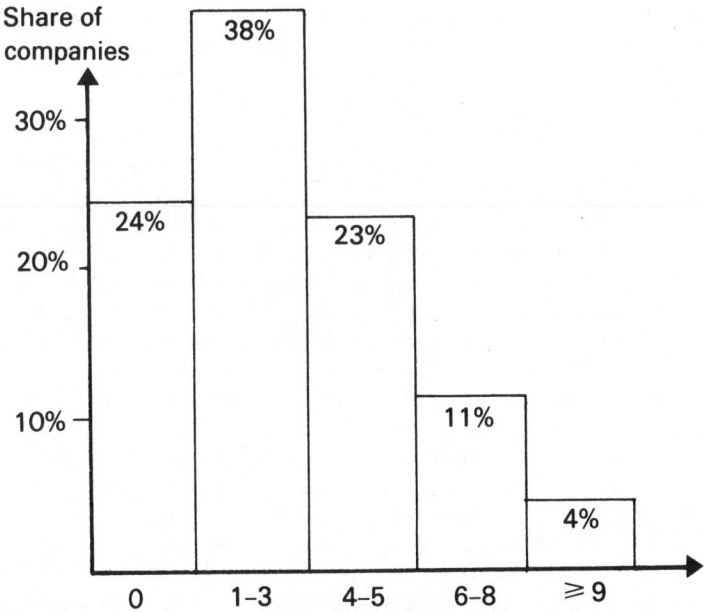

Number of development relationships with horizontal units

Figure 6.9 Number of horizontal technological development relationships

Resource structure

The effect of the technological dimension on the propensity to co-operate

In terms of technology, manufacturing companies can differ enormously. A company's production operations may be more or less advanced, and it may be making more or less sophisticated products. In trying to find relevant technological dimensions we can start with production and products. In the first instance these should define internal corporate characteristics, but at the same time they also automatically and inevitably reflect parts of the corporate environment.

Production technology can be described in a variety of ways. Here I have chosen a common economic approach, and have described it in terms of capital intensity. This is an expression of the amount the company has invested in capital assets (in advanced modern equipment) in relation to its number of employees or its investment in other assets. I have measured the capital intensity as the book value of capital assets divided by the balance-sheet total. The greater the investment in capital assets in relation to investment in current assets, the higher the capital intensity. It is assumed that the importance of production technology to the individual company is directly related to the company's capital intensity. As production technology becomes more important, links with external units should also increase in importance, which in turn should have a positive effect on the propensity to co-operate.

Products can involve different levels of technology, that is they can be refined to a greater or lesser extent. We measured the technological content of products by dividing the sales price by the 'weight' of the products, that is the sales price per kilo.[10] The more expensive the product in relation to its weight, the higher the technological content. For this dimension, too, we expected a positive correlation with the propensity to co-operate since the technological dimension becomes more important as the technological content increases, and this in turn should encourage the propensity to co-operate.

The results regarding capital intensity are shown in Table 6.16 and regarding the technological content of the products in Table 6.17. In the case of capital intensity the cross-classification shows a clear correlation between external relative share and capital intensity, that is the relationship we expected is confirmed by the results.[11] The importance to the company of the production technology is matched by an increasingly extroverted approach to technological development. In other words, there seems to be a correlation between internal corporate characteristics and the company's relationships with external partners. The high capital intensity makes close technological co-ordination necessary with suppliers and customers. Networks that are dominated by capital-intensive units should thus be characterized by close relationships among the member companies, and consequently by a certain rigidity in structure.

Table 6.16 Capital intensity and relative share of external technological development (number of companies)

External share		Capital intensity					Total
		0–0.19	0.2–0.29	0.3–0.39	0.4–0.49	0.5	
Low:	0–29%	8	14	6	3	7	38
Medium:	30–64%	6	7	11	3	5	32
High	65–100%	4	8	8	5	11	36
Total		18	29	25	11	23	106*

Note:

*Seventeen missing values.

This does not necessarily mean that the development work is more important; it could also indicate that higher capital intensity generates a certain rigidity in production, which in turn calls for close technological co-ordination with both suppliers and customers.

The technological content of the product, on the other hand, gives us a U-shaped relationship.[12] Companies whose products exhibit a high or a low technology content have a lower average share of external development than companies occupying a middle position. It is interesting to note that this middle group also exhibits the greatest variation in the relative external share. Table 6.17 demonstrates clearly that companies do not only collaborate on advanced products. Relatively speaking, there are more companies with a high external share in the 'simplest' product group than there are in the 'most complicated' product group. Thus the simple positive correlation that we expected between the propensity to co-operate and technological content did not appear. Instead, it seems that collaboration flourishes in a middle area where the technology is complicated enough to provide a number of development opportunities, but still simple enough to be defined and divided into separate projects. However, the most important lesson here is that there are always possibilities for collaboration, regardless of the technological level of the product.

The technological content can also act as an indicator of the company's main network affiliations. The greater the technological content, the more likely it is that the relevant networks will exhibit technological dimensions, that is, will have a more noticeable element

Resource structure

Table 6.17 Technological content of the product and relative share of external technological development (number of companies)

External share		Product sales price per kilogram (in kronor)			Total
		0.1–9	10–49	≥ 50	
Low:	0–29%	11	15	13	39
Medium:	30–64%	12	8	13	33
High:	65–100%	9	20	10	39
Total		32	43	36	111*

Note:

*Twelve missing values.

of 'high-tech'. If we re-examine the results in Table 6.17 in this light, we could say that the middle group is involved in networks characterized by technology and which are also carefully structured to make it easy to find suitable collaboration partners. At a lower level of technological content the networks are still characterized by technological dimensions, but these are probably regarded as more obvious. At a higher level of technological content the networks are less structured, that is, collaboration is presumably even more necessary than in the middle position, but it is more difficult to find suitable long-term partners.

Thus there are clear relationships between the technological nature of the company and corporate behaviour *vis-à-vis* external partners on technological questions. But as Table 6.17 in particular has shown, in all positions the relationships that obtain are not simple and direct but are very much more complex.

Summary

Horizontal collaboration partners include all types of knowledge-producing units such as universities and research institutes, and for many companies should represent an important source of development. But this group as a whole is exploited as collaboration partners far less than customers and suppliers. For example, on an average, they were used as partners by our companies only half as often as customers.

It was impossible to identify any network structure, as the horizontal units represent such diverse institutions. Instead, we described the corporate technology in terms of production plant and products, and were then able to analyse the correlations between these and the propensity to co-operate. Capital intensity, which we chose as a characteristic of production technology, has a clearly positive effect on the propensity to co-operate. Companies with higher investment in plant are thus more predisposed to co-operate. We suggested that this could be because it is then more important to integrate development activities with outside units, or because in this case the technology is more inflexible and all changes consequently have to be made in close collaboration with various external partners.

On the other hand, the technological content of the products, measured as the price per kilo, produced a reverse U-shaped relationship with the propensity to co-operate. The propensity was greater in the middle group and lower in both extreme groups. It was interesting to see that the relative external share may even be greater in the case of products that are regarded as technically simple than in the case of more advanced products.

Technological collaboration partners – a summary

The companies in our study each had an average of about ten collaboration partners of various kinds. The dispersion is shown in Figure 6.10. Almost half of the collaboration partners are customers; suppliers are the next most popular group, while development relationships with horizontal units amount to barely half the number of relationships with customers.

Figure 6.10 provides a good illustration of two fundamental problems for management in connection with collaborative relationships. The first concerns the distribution among different types of partners, and the choice of specific partners. In all directions there may be several hundred possible options. Although I do not postulate a rational method of selection, assuming rather that the collaborative relationships have evolved gradually out of previous events, there is none the less always good reason and abundant opportunity to question and influence the established structure.

The second question concerns the co-cordination between different development relationships and between these relationships and internal events. The different types of partners are often handled by different departments in the company, and this adds a further complication to the task of co-ordination. In Chapter 10 I shall return to both these management problems.

Resource structure

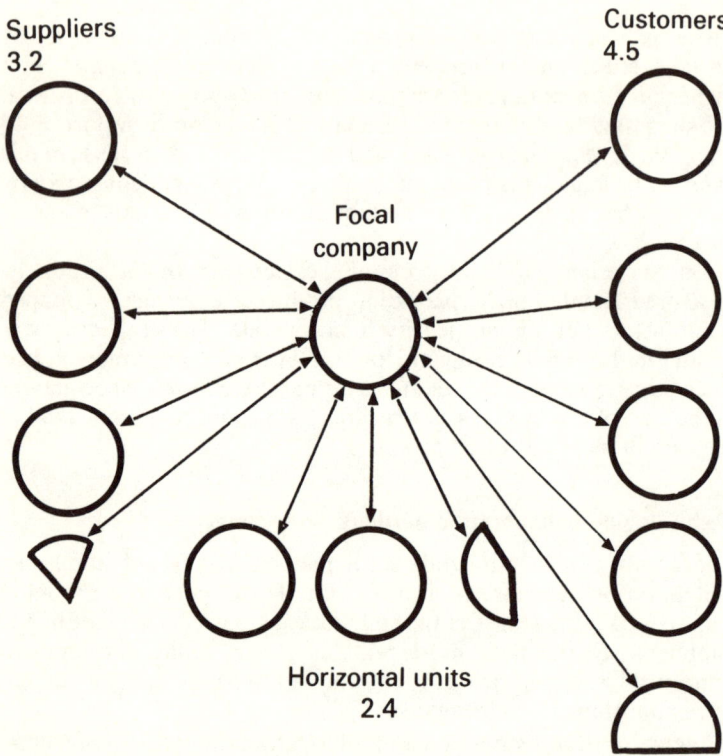

Figure 6.10 Corporate collaboration partners in technological development

Notes

1. A chi-square test gives a non-significant result at the 0.05 level.
2. A chi-square test gives, however, a non-significant result at the 0.05 level.
3. $r = -0.087$, $p = 0.361$.
4. $r = 0.156$, $p = 0.101$.
5. A chi-square test gives a significant result at the 0.06 level.
6. $r = 0.142$, $p = 0.135$.
7. For a more extensive discussion of personal networks see Hamfelt and Lindberg (1987) and Allen (1977).
8. $r = -0.181$, $p = 0.057$.
9. A chi-square test gives a significant result at the 0.05 level.
10. This way of estimating the technology content has earlier been used by Hörnell *et al.* (1973).
11. $r = 0.16$, $p = 0.098$.
12. $r = -0.113$, $p = 0.234$. A chi-square test gives a non-significant result at the 0.05 level.

Chapter seven

Corporate activities and network behaviour

In our introductory theoretical discussion, we described the network as an activity system in which exchanges serve to combine production activities in chains of varying length. An obvious assumption is that the different activities in such chains are mutually interdependent. Thus the propensity to co-operate on technological questions should be directly dependent on the nature of other activities being performed by the companies in the network. All these activities can be combined in various ways to form activity groups.

Since our starting point here is the propensity to co-operate on technological matters, we can look first at activities which dominate the technological dimension or which in a general way provide the framework for the exchange. A company's technical production activities are obviously an interesting activity group in our present context. Each relationship, and thus various collaborations as well, can be regarded as a link between different technical activity systems, and can thus be expected to reflect certain characteristics of these systems. Another interesting group is made up of all the corporate development activities. Collaborative undertakings represent a subgroup of these, and must depend directly on the nature of the activity group as a whole. Finally there are 'organizing' activities which link together diverse activities performed in different parts of the company, and which provide the institutional framework for various collaborations. We shall be examining each of these activity groups below, but first mention must be made of an important constraint on the empirical study. For reasons connected with the design and scope of the study, we were able to explore the activities of only one party, namely those of the companies in our sample. Obviously, we would have liked to examine the way their external partners designed their activities, as these were just as important as the activities of 'our' companies, but it was not possible for us to collect the relevant data.

Corporate activities

Development activities

An important basis for a company's technological collaboration with different partners is the totality of its own development activities. These determine what the company can do, and how attractive it will appear to potential partners. In Chapter 5 we discussed the propensity to co-operate and its relation to corporate results, and we noted the importance of balance between internal activities and external co-operation. If the company increases its total development investment, it can also increase its commitment to collaboration, without necessarily altering this balance. This does not mean, however, that the relative collaborative share increases. In fact the opposite probably applies. Our initial hypothesis is therefore that the more a company invests in technological development, the more it performs under its own auspices. But if the frequency of collaborative activities as a whole increases, then the number of collaborative partners will probably increase as well.

Table 7.1 shows the correlation between total corporate investment in technological development and the propensity to co-operate. The results do not support our expectations at all. There is certainly a relatively even dispersion of the propensity to co-operate in relation to total development investment, but the tendency is the opposite of what we had assumed. The relative propensity to co-operate rises with an increase in total investment in technological development. The increase in the weighted average between the three values identified for total investment is as much as 10 per cent. In other words, companies with a high overall level of development investment may have a higher relative share of external collaboration than companies whose investments are more modest. Perhaps the explanation is that companies investing a great deal in technological development raise their level of activity and more often receive invitations to collaborate with other companies. Thus companies with a higher level of activity would also have more collaborative partners than companies with a lower level of activity. However, the empirical study reveals no such tendency; on the contrary, the difference in the number of active partners is less than the difference in the size of the external share. On average the number of partners only varies by around 5 per cent, that is the companies which make the biggest investments have an average of 5 per cent more partners than those which invest less. And again, if we look at the distribution between different types of partner, we find no big difference between companies that can be related to the size of their total investments. The higher level of overall external activity in the development-intensive companies involves much the same number of partners as those with which other companies collaborate. In other words, it is primarily the intensity of the collaboration that is affected, and not the number of collaborative partners.

Table 7.1 Total development investment and the relative share of external collaboration (number of companies)

External share		Total development investment measured as a percentage of work-years			Total
		< 2	2–4	> 4	
Low:	0–29%	15	21	6	42
Medium:	30–64%	15	20	3	38
High:	65–100%	10	23	6	39
Total		40	64	15	119*
Weighted average for relative share of external collaboration		45.9%	49.6%	49.0%	48.3%

Note:

*Four missing values.

It is interesting to note in Table 7.1 that the relative external share is very evenly dispersed over all three levels of total development investment. In other words, there seems to be a considerable degree of freedom for the individual company in regard to the relationship between external collaboration and total investment. Let us now go one step further, and see whether the nature of the total investment affects the propensity to co-operate.

To simplify, we can say that the total investment of the individual company can be divided between product development and process development. A closely related factor then is the extent to which development is geared to cost-minimization. Both product and process development can be conducted in collaboration with external partners. However, we have already noted that customers are the most frequent partners, which would suggest a higher external share in product development. The reverse should be true of cost-minimization, which can be regarded as a more introverted activity than seeking new developments, quality improvements, and so on.

Our results suggest that a certain difference does exist in the propensity to co-operate, depending on variations in these two dimensions, and the trend coincides with our expectations. Up to a point companies with a relatively high level of investment in product

Corporate activities

development also show a greater propensity to co-operate, and companies with a bigger element of cost-minimization show a lower propensity. But the differences are slight, and this supports our earlier conclusion that companies have a good deal of freedom to choose their level of collaboration, regardless of the nature of their total development investment.

Production activities

Customer and supplier relationships can be regarded as an extension of corporate production activities but in opposing directions. It therefore seems natural to assume that the technological content of these relationships, for instance in terms of the propensity to collaborate, should be influenced by the nature of the production activities. We have already touched lightly on this point in our discussion of technological structure above. We mentioned technological content and capital intensity, for example, both of which are important aspects of production activities. We can now look at three other aspects. First, the general production technology which we described by classifying the companies in our empirical study as being engaged in unit production, small-batch production, mass production, or process production. We also found companies whose technology consisted of combinations of these types. We assumed that the production technology would prove less flexible and more controlled the closer we came to process production. Here input and output are both determined more directly by the technology. Thus if a company wants to introduce some change connected with technological development, it is also necessary to make some change in relations with suppliers or customers. We could then expect to find a higher relative share of external collaboration in these companies.

This hypothesis receives some support from our data.[2] Companies involved in process production, or employing a combination of production technologies, have a higher relative share of external collaboration than companies in the other three categories. However, there is very little difference between these other three. It is also interesting to note that when the amount of collaboration is affected by the type of production technology, it is mainly supplier relationships that are involved, while there is very little effect on the situation *vis-à-vis* customers or horizontal units. Thus production activities as described in these terms have relatively little effect on the propensity to co-operate.

Production activities can be steered by customer orders or by stock requirements (production for stock). In the first case a specific order and a specific customer determines the design of production, while in the second case it is determined by the company itself, in light of an assessment of the sales prospects. We would expect to find that companies engaged in customer-led production have a higher share of collaborative projects than those engaged in production for stock. Our results support this hypothesis.[3] The proportion of customer-led production plays an important part, as can be seen in Table 7.2. The correlation is clear and strong: the higher the proportion of customer-led production, the higher the propensity to co-operate.

Table 7.2 Relative share of external technological development in relation to proportion of customer-led production (percentage of companies)

External share		Proportion of customer-led production			Total
		0–9%	10–74%	75–100%	
Low:	0–29%	11	10	13	34
Medium:	30–64%	6	5	20	31
High:	65–100%	4	7	24	35
Total		21	22	57	100
Weighted average for external share		35%	43%	54%	48%

There is obviously a close connection between a company's general situation *vis-à-vis* its environment and the way it collaborates with external partners on technological matters. If a company has a generally extroverted stance – as companies producing for customer orders presumably have – this will also affect the way technological questions are handled. Consequently, the problem for this type of company is more a question of limiting and targeting the technological development work. Companies with a high proportion of production for stock, on the other hand, may have to make a conscious effort to initiate collaborative projects.

The amount of value added is a third aspect of production that should be interesting in connection with the propensity to co-operate. A substantial amount of value added means that a company acts as a link between suppliers and customers which, purely functionally, are farther away from one another than in those cases of a lower value added. This could be expected to mean that the interface with both customers and suppliers would be more important in a development perspective, with a consequently greater propensity to co-operate. However, there is no support for this supposition in our data. We cannot identify any direct correlation in our data between the propensity to co-operate and the amount of value added.[4]

Organizing activities

We are now becoming accustomed to the great variety of activities that fills the corporate world. And every activity, on every separate occasion, takes place within certain given resource limits: plant, machines, and personnel. In order that the resources should be efficiently utilized, various organizing activities are required. There has to be standardization and specialization, so that the company can achieve advantages of volume and scale. Altogether a great many actions are required, involving not only internal departments but also external units, to make it possible to co-ordinate the countless activities.

Technology is one of the dimensions which has to be organized in this way. Technological changes occurring in one part of a company have to be co-ordinated with changes occurring elsewhere in the company, or with changes in customers and/or suppliers. Collaborative projects can provide a fruitful means of achieving this.

Organizing activities may be directed at external units such as distributors and agents, at independent internal units such as subsidiaries (sales, manufacturing and/or development), and at integrated internal departments such as production, purchasing, development, and marketing. Organizational changes may involve great leaps (major reorganizations) or small successive adjustments. The organizing activities reflect external changes in various ways, and we could therefore expect them to be closely related to technological development in the relevant networks.

Internal organizing activities constitute such a large and complex problem as to call for a special study; in the present investigation, unfortunately, we have had no opportunity to tackle such a study even in the form of an overview. We concentrated instead on organizing activities in the borderland between a company and its customers. Customer relations have to be handled and co-ordinated in various ways, and a variety of internal or external units can be used for the purpose. First, purely external units such as distributors or agents may be responsible for customer relations within a particular geographical area. Second, more or less independent internal units such as subsidiaries may have a similar responsibility; and third, customer relations may be handled directly by the company concerned.

We will limit the analysis here to examining the use of distributors in the domestic and export markets. Distributors do not represent the sales company; they buy products from the company outright, and resell them to different kinds of final customer. Table 7.3 shows the extent to which the companies in our sample made use of distributors. In the domestic market roughly one-third of the companies used distributors a great deal, while in the case of exports the figure was 50 per cent. Let us now explore the implications of this in connection with technological collaboration with customers.

Table 7.3 Use of distributors (percentage of companies)

Distributors' proportion of sales in the particular region		Domestic	Export
	0%	42%	39%
Low:	1–39%	20%	9%
Medium:	40–74%	10%	4%
High:	75–100%	28%	47%
Total		100%	100%

By using distributors a company can reduce the cost of customer relations, as the distributor fulfils a merchandizing function. At the same time the distance between the company and its final customer is considerably increased, since most contacts are now indirect. We would expect this to result in fewer technological collaborative relationships with customers, which in turn would mean a lower relative share of external technological development activities. Could not the distributors take on the role of collaborative partners? Up to a point they can, in so far as they pass on customer reactions and views to the producer. But they can never altogether take over the customers' role, since they belong essentially to another technological world; they have other technical problems and opportunities associated with their distributor role.

Our data confirm this line of reasoning clearly and unequivocally.[5] We can see in Table 7.4 that the number of development relationships with the ten largest customers (distributors have been counted as customers in this context) falls steeply as the distributors' share of sales rises.

The 37 companies that mainly use distributors have noticeably few development relationships with customers. If the analysis is extended to include other customers as well, as the ten biggest, the conclusion remains the same (see Table 7.5). Even here there is a marked difference between companies that do not use distributors and those that do.

Corporate activities

Table 7.4 Correlation between use of distributors and technological development in customer relationships (percentage of companies in each category)

Distributors' share of corporate sales in Sweden		Proportion of the 10 largest customer relationships with co-operative development			Total	Number of companies
		0–2	3–6	7–10		
Low:	0–29%	37%	10%	53%	100%	70
Medium:	30–59%	30%	20%	50%	100%	10
High:	60–100%	73%	19%	8%	100%	37

Thus the use of distributors appears to result more or less automatically in few technologial collaborative relationships with customers. In this analysis we have limited ourselves to distributors on the Swedish market, but the result would have been exactly the same if we had included the export market.

Table 7.5 Development relationships with small customers and the use of distributors (percentage of companies in each category)

Distributors' share of sales		Frequency of technological development collaboration			Total	Number of companies
		Always or often	Sometimes	Seldom or never		
Low:	0–29%	22%	18%	60%	100%	72
Medium:	30–59%	9%	18%	73%	100%	11
High:	60–100%	9%	20%	71%	100%	35

Activities and the propensity to co-operate – a summary

In the companies studied we found a clear correlation between each of the three main types of activity identified and the propensity to collaborate. Unexpectedly, the overall size of development investment proved to have a positive effect on the relative share of external collaboration. The distribution of the total investment was also important, and we found that the larger the proportion devoted to product development, the more a company invested in external projects. Despite these correlations, our main impression is that the companies enjoy great freedom in forming their collaborative profile, more or less independent of the nature of their total investment in development.

The same conclusion cannot be drawn when it comes to production activities. Here the relationships, and consequently the restrictions, are stronger. In the first instance this applies to whether production is geared to customer orders or stock requirements. One would naturally associate the first case with a high propensity to collaborate. In other words, companies involved in production for stock must work much harder and strive consciously to achieve the same relative share of external development investment that the customer-led company more or less automatically acquires.

In the case of organizing activities, we found that the introduction of an intermediate level between customer and company, such as a distributor, agent, or subsidiary, led directly to a reduction in the propensity to co-operate. Using an intermediate level seems to be like putting up a wall: great efforts are needed to break through this wall, and only in a few cases is such a break achieved.

Thus according to our results there is a big difference in the propensity to co-operate between, on the one hand, companies that overall invest a great deal in development, gear their production to customer orders, and are in direct contact with their customers, and, on the other, companies that invest much less, engage in production for stock, and sell through distributors. The latter companies are automatically introverted, while the former are just as naturally inclined to be extroverted. In the introverted company management faces certain problems, largely due to the fact that the company makes too little use for its own benefit of external technological resources and changes in the environment. The management of the extroverted company has problems of another sort. Here it is more a question of trying to hold the company together by finding a particular line of development among all the suggestions and demands crowding in on the company from all directions.

Notes

1   $r = 0.037, p = 0.69$.
2   $r = 0.300, p = 0.01$.
3   $r = 0.245, p = 0.01$.
4   $r = -0.011, p = 0.91$.
5   A chi-square test gives significant results at the 0.05 level.

Chapter eight

Individual development relationships – characteristics of the partners

A relationship arises as a result of interaction between two companies, or rather a series of interactions during which the behaviour of both parties affects the sequence of events. We have already found that corporate behaviour in these relationships depends in part at least on various characteristics of internal and external situations. Thus the interaction is likely to be influenced not only by characteristics of the two parties *per se*, but also (and probably even more) by the way these characteristics interact with one another. Is it true, for instance, that 'birds of a feather flock together'? And, in a related question, what dimensions are particularly important? Three characteristics which we shall be discussing further in the first section below are geographical location, volume, and familiarity. We will analyse the importance of collaborative partners in these terms, as it is perceived by the companies in our study.

The first section pursues the subject explored in previous chapters, namely the propensity to co-operate and various factors that affect it. The difference is that we are now discussing relative conditions, and the empirical data consist largely of important individual relationships. With the exception of the case of geographical location the data refer not to all collaborative relationships but only to the most important (a maximum of three of each type, i.e. the relationships for which we received more specific information in response to questions 62–97, 167–199, and 253–282 in the Appendix questionnaire).[1]

The subsequent sections deviate from our previous pattern and consist instead of a deeper analysis of collaboration as such. For example, we would expect an evolving collaborative process to be affected by the nature of the interaction – the forms of co-operation perhaps, or the contact structure. The process may be more or less self-generating for instance, depending on whether the technological problems or opportunities that emerge at one stage are used as a starting point for the next. Here we must examine influential dimensions such as the degree of formalization, the number of people involved, and the frequency of contacts. These dimensions will be examined more closely in the second and third sections.

In rather the same way we would expect results previously achieved in the relationship and the expectations for the future to be important determining factors. We will examine this aspect in the fourth section below. Finally, in the fifth section we will seek an overview of the individual development relationships.

Characteristics of collaborative partners

Our study included a geographical dimension. In theory, geographical proximity should be a considerable advantage to a company that wants to collaborate technologically with another. Proximity makes personal contact both easier and cheaper, and this should be a valuable advantage to technological collaborations demanding the extensive and intensive exchange of information. Thus geographical proximity should always be a positive factor. The only argument against this is the increasing specialization which companies have been experiencing throughout the present century, and the consequent need for co-operation with highly specialized partners. The competence, knowledge, and interest of the collaborative partner may so outweigh geographical proximity in importance as to render it of marginal interest only.

We can return now to the empirical study. Table 8.1 outlines the three types of collaborative partner which we have identified, and several interesting points emerge. Comparing the propensity to co-operate of different partners in relation to one another, we find that – compared with other partners among suppliers or customers – horizontal partners are more often located within the same region. Thus local/regional networks seem to be more important when there is no ongoing commercial exchange to provide a basis for collaboration. Or, conversely, ongoing commercial exchange is a good way of overcoming the problems arising from greater distance.

As far as customers are concerned, however, the customers in the local region are more important in collaborative terms than they are in terms of sales volume. They account for 17 per cent of development relationships, but for only 12 per cent of volume. More distant customers (foreign customers) account for a lower proportion of technological relations than their volume share would suggest, namely 42 per cent of volume but only 18 per cent of development relationships.

In regard to suppliers, it is impossible to estimate the situation in

Individual development relationships

Table 8.1 Geographical dispersion of different types of partner and of total sales or purchases

Geographical location in relation to company studied	Share of development relationships customers	Share of development relationships, suppliers	Share of development relationships, horizontal units	Share of sales	Share of purchases
Domestic					
same region	17%	16%	25%	12% ⎫	
rest of Sweden	65%	62%	62%	56% ⎬	80%
Foreign					
the north	4%	6%	2%	16%	6%
Europe	11%	12%	6%	11%	11%
other	3%	4%	5%	5%	3%
Total	100%	100%	100%	100%	100%

the region with any certainty. However, Table 8.1 shows that the foreign suppliers account for a higher proportion of development relationships (22 per cent) than of volume (20 per cent). Geographical distance seems to be less of an obstacle on the buying than on the sales side. At the same time this result also suggests that the notion of foreign suppliers being popular because they are cheaper should be promptly replaced by the recognition that they are used because they are technologically more competent.

There is an interesting difference in the case of the Nordic countries. Nordic customers account for slightly over one-third of exports but only about 20 per cent of foreign collaborative relationships. On the supplier side, however, the Nordic countries account for a share of development relationships that easily matches their volume share.

The geographical location of partners thus seems to be an important factor, but one whose effect varies for different types of partner. On the customer side priority is given to nearby partners, but in collaborative activities with foreign customers the Nordic countries are often neglected. Within the local region horizontal collaborative partners clearly have priority. On the supplier side, however, the opposite is true, and foreign (including Nordic) suppliers account for a share of development collaboration easily on a par with their volume.

Another important factor which determines a company's choice of collaborative partner is the relative importance of the partners to one

Individual development relationships

another. If two companies are important to one another in terms of volume, they are also likely to be important in a technological context. It is always important for a seller to keep up with the technological development of major customers, and for a buyer the technological development of volume products can loom particularly large.

Our empirical data support this hypothesis regarding the importance of volume. A quarter of the customers with which our companies collaborated accounted for more than 50 per cent of the volume in the relevant product group. Another quarter accounted for between 20 and 50 per cent. Only a fifth accounted for less than 5 per cent. The picture is similar on the supplier side. Half the suppliers with which our companies collaborated accounted for more than 5 per cent of the total volume purchased, a third accounted for more than 10 per cent, and about an eighth for more than 25 per cent of the total volume. Thus there is no doubt at all that the most intensive collaboration occurs in relationships that are important in volume terms. But which way does the correlation point? Does volume come first and collaboration follow, or vice versa? We should get some information about this from our analysis of the next dimension, which concerns the duration of the particular collaborative relationship. What we are really trying to find out is whether the company generally collaborates with established customers and suppliers, or whether it uses technical collaboration in order to launch a relationship. If the latter, then the technical collaboration precedes the volume, but if the former situation obtains, then volume presumably comes first and collaboration follows.

Table 8.2 shows the age distribution of the development relationships in the companies studied according to their duration. The result is extremely clear. Two-thirds of all customer and supplier relationships which include an element of technological development have lasted for longer than five years, and almost half of them for longer than 15. If we examine the less-than-five-year group, we find that only one-third of these relationships have lasted less than two years (i.e. 11 per cent of the total). In other words, it is quite clear that on the customer and supplier sides companies collaborate in the first instance with well-established and familiar partners. The horizontal development relationships deviate somewhat from this pattern, however. Here over half of all the development relationships are less than five years old. In a way this is quite natural, since unlike supplier and customer relationships, horizontal relationships do not depend on the day-to-day flow of goods and payments. Even among the horizontal relationships, however, there are a good many in which the partners have been in contact with one another for a fairly long time.

Individual development relationships

Table 8.2 Duration of development relationships in relation to type of partner (percentage of relationships)

Duration (years)	Customers	Suppliers	Horizontal units
0–4	36%	28%	55%
5–14	30%	41%	29%
≥ 15	33%	29%	15%
Weighted average	13 years	13 years	8 years

The results for geographical location, volume, and familarity can be explained primarily in one of two ways. Either collaboration on technological development evolves because important and well-known partners on the domestic market, or customers on major large export markets, can more or less force a company into such development collaboration; or the company can utilize its technological development collaboration with the same partner to improve its position *vis-à-vis* the partner, either directly or indirectly by improving its own competence as a result of the technological collaboration. The final outcome is probably affected by the status of the partner and the intentions of the company in question. On the other hand, it seems clear that collaboration on technological development is seldom used as a way of acquiring new customers or suppliers. Thus collaboration seems to call for earlier acquaintance between the two partners, and consequently also for a certain volume before collaboration commences. At later stages there are presumably positive or negative spirals: successful collaboration leads to an increase in volume, which generates more collaboration and so on, or vice versa.

Form of co-operation and type of development collaboration

Co-operation can occur in various ways. For example, the arrangement can be more or less formalized. At one extreme collaboration might be completely informal, possibly based on an ongoing commercial relationship in which the technological collaboration is only part of a much larger whole. At the other extreme the collaboration may be extremely formalized, and at least from the legal point of view, defined as a relation in itself. In our empirical study we generally expected to find several formalized relationships with horizontal partners, where there is no continual flow of commercial activities. Further, we expected that at least the bigger and more extensive collaborative undertakings would be formalized.

However, the results in Table 8.3 do not suggest such a pattern. By far the most common form of co-operation is the informal version, either in the shape of ongoing commercial relations or in some other form (mainly for horizontal relationships). Even where there is a formal arrangement (e.g. an annual agreement), these are not usually specific to technological development but refer to the whole relationship between the parties (i.e. volumes, prices, and so on). We could conclude from this that technological development activities represent an integral part of overall corporate operations which cannot easily be singled out, and which are also used for purposes other than the purely developmental. Another conclusion could be that most technological collaborative projects are so diverse in nature that it is difficult to define them in formalized agreements. It would therefore be interesting to look a little more closely at the types of different development collaborations.

Table 8.3 Form of co-operation (percentage of relationships)

Form of co-operation	Customers		Suppliers		Horizontal units	
Formalized relationships						
Annual agreement	20%		11%		1%	
Long-term agreement (more than 1 year)	13%		8%		8%	
Joint company (or similar)	2%	35%	2%	21%	11%	20%
Non-formalized relationships						
Part of ongoing relationship	51%		67%		41%	
Other informal design	14%	65%	12%	79%	39%	80%
Total		100%		100%		100%

Collaborations can vary enormously in content, as seen in Table 8.4. We set as a lower limit 'the mutual exchange of technological information'. In these cases the partners each conduct their own technological development, but close co-ordination is established in the mutual exchange of information. In order to count as collaboration in our study, 'exchange of information' had to have had some effect on technological development. Tests and trials constitute another possible ingredient in collaboration, and they also help to shape the outcome of the mutual exchange of information. At the next level, one partner conducts a particular project on the

Individual development relationships

other partner's account. At a yet higher level, two partners establish a joint project group which then carries out development work. And finally, there is the kind of long-term technological collaboration in which the two parties work together on development issues over a very long period. These five levels were intended to form a scale from simple co-operation involving very little investment to complex collaborative arrangements accompanied by major investments. However, when we came to collect data we found that even collaborations on special projects, or those consisting mainly of tests and trials, could be fairly complex and often required heavy investment. Thus the five levels should be seen simply as categories describing the nature of the collaboration. The results clearly indicate the great diversity of content in the relationships.

Table 8.4 Type of development co-operation (percentage of relationships)

Type of co-operation	Total	Customers	Suppliers	Horizontal units
Mutual exchange of technological information	22%	18%	24%	24%
Tests, trials, etc.	18%	17%	21%	16%
Special technological project (either side)	26%	34%	27%	14%
Joint development activities including project group	14%	14%	10%	19%
Long-term technological collaboration	15%	14%	12%	18%
Other (including licence agreements)	5%	3%	6%	9%
Total	100%	100%	100%	100%

One-fifth of the cases are primarily concerned with the mutual exchange of information, and roughly the same number consist of tests and various types of trials. Special projects account for about a quarter of the cases, and joint development activities, including long-term technological collaboration, account for almost a third of all the development relationships. As regards any variation between different types of relationship, the differences are small. Only in the case of special technological projects do the horizontal relationship deviate slightly. The figures in Table 8.4 provide powerful support for our earlier conclusion that the contents of collaborative relationships are so diverse that it is difficult to contain them within formalized agreements. When we add to this the long duration of most relationships, it becomes obvious that this collaboration is closely associated with other dimensions in a relationship, which should thus always be considered in their entirety.

Personal contacts

A crucial and significant characteristic of the interaction consists of the personal contacts it involves. These make possible a freer and more lively exchange of ideas and suggestions than any written document can promote. People on both sides can gradually build up confidence and trust in one another, and in this way an important social element enters the interactions.[1]

Earlier studies (reviewed, for instance, in Turnbull and Valla 1986) have described the wide-ranging and lively contacts in important customer and/or supplier relationships. Thus we also expected to find the same picture in the technological collaborative relationships in our study, since the technological content presumably involved technology of various kinds in the relationships. And our results did indeed illustrate a wide and lively range of contacts (see Tables 8.5 and 8.6). An average of three to seven people from the companies concerned took part in the contacts and personal meetings occurred every month or every other month on average. There were, however, many variations. For example, in 20 per cent of the customer relationships at least ten people were involved on each side, and in 17 per cent of these relationships contact was made at least once a week.

Table 8.5 Number of people having direct contact with partner in different development relationships

Number of people	Customer relationships		Supplier relationships		Horizontal relationships	
	Studied company	Customer	Studied company	Supplier	Studied company	Horizontal unit
1	11%	7%	13%	14%	20%	26%
2–3	45%	39%	54%	53%	54%	33%
4–6	24%	26%	24%	22%	17%	19%
7–9	4%	7%	7%	5%	6%	7%
10–14	8%	10%	1%	5%	3%	5%
≥ 15	9%	11%	2%	1%	1%	11%
Average no. persons	5.5	6.4	3.8	3.8	3.4	5.3

Individual development relationships

Table 8.6 Frequency of contact (personal meetings)

Frequency	Customer	Supplier	Horizontal units
Once every 3 months at most	45%	60%	50%
At least twice every 3 months, but at most once a fortnight	38%	33%	35%
At least once a week	17%	7%	15%
Total	100%	100%	100%

In regard to the dispersion among the studied companies and their different types of partner, it seems that in two of the cases more people were engaged in the partner company than in 'our' company. This applies particularly to horizontal relationships, but also to some extent to customers.

These lively contacts suggest a number of important problems which we have not been able to explore here, but which could provide fruitful subjects for future research. First, the frequency of the contacts indicates something of the cost of these relationships. Thus companies must always be on the look-out for efficient ways of managing the communications aspect of their collaborations. In view of the technological advances that have already occurred and are still occuring in the field of communications, several interesting lines of development are suggested here.

Second, with the large numbers of people involved it is obviously important that relationships be co-ordinated, to avoid misunderstandings, contradictory signals, and so on. The difficulties are not alleviated by the fact that the people taking part often represent different departments in the company, and the organization of the relationships thus becomes something of a key issue.

Third, the large numbers of people involved suggest that companies must be influencing one another in a wide-ranging and probably uncontrolled manner, and that there are certainly a great many loose threads lying about. The network contains innumerable slender strands, some of which are used for solving problems or exploiting opportunities while others remain dormant. Thus in all these forms there must be great potential for technological developments that never materialize.

Results of the relationships

The results of the development activities can vary both in form and in the way they are distributed among the parties concerned. The result may also consist of one or several improvements, and these can vary in size. Since we had been exclusively studying relationships that were both reciprocal and of long duration, we expected there to be results positive to both parties, often consisting of more than one improvement. And our results tend to support this expectation. Table 8.7 shows the results for the companies studied and for their different types of partner. In most cases the result led to positive technological improvements in the companies interviewed (between 85 and 90 per cent). However, the same companies felt that their partners had gained less from the collaboration, at least in the case of suppliers and horizontal units where it was felt that a third of the partners had gained no concrete advantage.

Table 8.7 Results of the collaboration

Type of result	Type of partner		
	Customers	Suppliers	Horizontal units
Nothing concrete	15%/16%*	8%/30%	16%/37%
A slight improvement	14%/11%	16%/18%	6%/10%
Several slight improvements	25%/25%	37%/22%	35%/24%
A major improvement	24%/27%	29%/20%	26%/15%
Several major improvements	22%/20%	9%/10%	16%/14%

Note:
*Figures on the left refer to the company studied and those on the right to the partner.

It is interesting to note that almost half the collaborations have resulted in several improvements, which suggests that collaboration may start for one purpose but subsequently lead to a variety of other projects.

Customer relationships are judged to be fairly well balanced, in that on average the partner gained as much from them as the studied company. In the case of suppliers and horizontal partners, this was not so. Here the companies studied were apparently more inclined to utilize their partners for their own technological development, without putting as much into it themselves. This result may depend on the realities of the situation, but it could also depend at least partly on different perceptions of customers and suppliers. Customers have to be managed, and it is important that they should gain something from the collaboration. Perhaps companies therefore try to see that customers do gain something from it, and even try to discover and point out all the improvements that may have resulted. But perhaps when it comes to suppliers and horizontal units, less effort is made; companies may feel these groups can take care of themselves.

We also asked our respondents what they expected from the different development relationships. Table 8.8 shows that there is usually a general expectation that 'several slight improvements' or 'several major improvements' will result. Thus the picture we outlined above is confirmed. The collaboration is not specific, but is of a more general kind; consequently, it is possible to expect a variety of technological improvements to result. An interesting similarity between Tables 8.7 and 8.8 is the predominance of 'several minor improvements' on the supplier side. On this point the suppliers differ markedly from the other partners.

Table 8.8 Expectations regarding the results of established collaborative relationships

Expectations	Customers	Suppliers	Horizontal
Nothing concrete	30%	24%	19%
A slight improvement	4%	13%	4%
Several slight improvements	36%	48%	39%
A major improvement	4%	5%	6%
Several major improvements	26%	10%	32%
Total	100%	100%	100%

Summary

In the course of this chapter a picture of technological collaboration as a natural part of total corporate behaviour has gradually emerged. First, we found that the typical partner is near at hand, important in terms of volume, and well established as a partner. Thus general social characteristics are more important than purely technological or knowledge attributes. The conclusion seems to be that companies enter into technological collaboration with units which they have come to know in other interactions, rather than interacting with units they would like to acquire as technological collaborative partners.

The picture of collaboration as an integral part of a broader relationship is reinforced by the great diversity of content and the high level of informality that we also identified in technological co-operation. Thus it is difficult to distinguish the collaboration from the relationship as a whole; rather it represents a complex part of a larger and even more complex whole. Obviously, there will thus be intimate and multifaceted dependencies between the technological collaboration and the relationship as a whole. And if we are to analyse and evaluate the technological collaboration, we will have to take these conditions into account.

The complexity of the situation emerged very clearly from our descriptions in the last two sections, which revealed the number and extent of the personal contacts, and showed that collaborative relationships often result in a variety of improvements on both sides.

Note

1 Total number of such relationships were 496; 172 were with suppliers, 168 with customers, and 156 with horizontal units.

Chapter nine

Technological co-operation and corporate relationships – summary and theoretical implications

Anyone involved in technological development knows that the technical solution to every problem will mean the birth of at least two new problems. Similarly, but on a more positive note, corporate relationships tend to open the way towards a variety of new solutions rather than tying the actors down to one – or at least so it seems when technological co-operation comes into the relationships. The commitment then possesses a dynamic element, and instead of having an inhibiting effect it provides fertile soil for change. Relationships which in themselves may be seen as a tie or an obstacle may in fact represent the grounds for technological development; without them, perhaps, no change would have been either necessary or feasible.

Our empirical results provide some support for this notion, but to be able to test it fully we would need further studies of a different kind. Even as a hypothesis, however, it represents such a fascinating idea that simply to have formulated it must be regarded as an important result of the present study. In the following summary of our other results I shall not repeat the conclusions already presented at the end of the relevant chapters, but will present the results a little differently. In the first part of the present chapter I shall concentrate on four points which can be regarded as the main results of the study. In the second part I shall examine the implications of these results for our theoretical network model. And in the two following chapters I shall explore the implications for business leaders and politicians, which will also mean adopting a slightly broader context than that of the study itself.

The results of the study can be summarized under four main headings. The first is the strategic importance to individual actors (companies) and to the network as a whole of the corporate relationships and the technological co-operation they include. Second, and closely related to this, is the nature of the relationships and collaborations in investment terms. The relationships which

Technological co-operation

have emerged over a long period, involving laborious and costly activity, represent one of the most important bases for technological co-operation. Third, we have learned something of the types of partner with which a company collaborates and of the way different types can be combined. Fourth, we look at the basic characteristics of the relationships and the associated effects.

Strategic implications

Our results have frequently confirmed the importance of technological co-operation to individual companies, and thus to the industrial system as a whole. There are several interrelated explanations for this. One concerns the often large proportion of technological development work which is conducted as part of some external collaborative project. Obviously this is something which is not easy to measure, but if we adopt a relatively strict definition of external co-operation we find that the proportion is around 50 per cent. Thus half the development work conducted by the 123 companies in our study takes place in collaboration with someone else. This is such a significant volume as to warrant treating technological co-operation as a strategic phenomenon. Companies are well aware of the need for a long-term overview of their own technological development work; it should be regarded as just as important, however, to maintain a long-term perspective on the external aspects of their development activities.

Another reason for emphasizing the strategic aspect is that technological co-operation is connected in various ways with certain internal characteristics of the company, for instance its production form and organizational design. For example, one company's production may be geared to customer orders while another uses distributors, and this will have a direct effect on their technological co-operation. Technological co-operation thus constitutes a natural element in overall customer management. But there are also obvious risks: companies who find it natural to collaborate with many customers may let their development situation become too fragmented, while those who find it natural to collaborate with customers on a small scale or not at all may be missing an important group of potential partners for technological co-operation. Various strategic expedients will be needed to prevent these natural obstacles or circumstances from becoming too compelling.

Technological collaboration is not only connected with internal corporate characteristics in this way, it also affects and is affected by the company's position in the network. For instance, we found a connection with the number of owners, the personnel structure, and the customer and supplier structures. Thus technological co-operation is related to many other important corporate dimensions, which should be borne in mind not only in designing the technological co-operation itself but in designing the other dimensions as well. In reviewing our results we spoke of the 'corporate identity' to describe the whole picture, so that 'identity' refers to the way others perceive the actor in question as well as the actor's own self-perception. Technological co-operation certainly represents one aspect of the identity, but it is also a means whereby the identity can be influenced. All the various relationships that were documented above can be regarded as opportunities for exerting influence by changing the technological co-operation, and as important conditions which can themselves be influenced, thereby creating fresh opportunities for technological co-operation.

A third reason for emphasizing the strategic importance of technological co-operation is that despite the relationships identified above, the individual company enjoys great freedom to design the technological co-operation to suit its own purposes. The co-operation profiles in our study, for example, varied greatly from one company to another, ranging from isolationist profiles to profiles consisting of many extensive collaborative activities. Much of this variety can be explained by internal characteristics and the company's network situation as described above, but much of it is obviously an expression of the company's own ambitions. In other words, there is plenty of room for manoeuvre in this dimension, which once again underlines its strategic importance. Our results showed, for instance, that companies which differ greatly in their relative investment in technological development may well have similar strategies when it comes to the proportion of this investment that goes into external collaboration. Thus some companies invest a great deal in technological development and maintain a high external share, while some companies invest only a small amount, but – such as it is – the investment goes towards external collaboration. The same applies if we look at levels of technological content. All in all, regardless of its internal characteristics or its position in the network, a company can design its co-operation profile with a fair amount of freedom.

The three reasons we have identified here to explain the strategic implications of technological collaboration – its volume, its links with various internal characteristics and the overall network situation, and its considerable freedom in designing the co-operation profile – together provide a clear picture of the strategic importance of the relationships in general and of the technological co-operation involved.

A fourth explanation, which will be discussed in the following section, concerns the heavy investment necessary to create the conditions for technological co-operation.

Investment-intensive relationships

Technological co-operation, as can be seen from our study, is generally based on well-established relationships – well-established, that is, in terms of the length of time they have lasted. The average among our companies was well over ten years. And every year a substantial number of work-hours had been spent to establish and/or maintain each such relationship. Time was spent on people getting to know one another, discussing various problems, finding solutions, convincing one another of the suitability of the solutions, and generally performing all sorts of practical actions. Thus a good many hours are invested in a relationship before it can provide a suitable basis for technological co-operation. And, what is worse, not all relationships actually result in technological co-operation. Nor, presumably, is it possible to forecast at an early stage the likelihood or appropriateness of any future collaboration. Thus a number of relationships have to be established just so that one of them can eventually lead to technological co-operation. If these costs are all allocated to the small number of collaborative relationships that are finally established, it can be said that a considerable investment has been made in each one of them.

Moreover, the value of the investment is uncertain. Many collaborative projects lead to fairly meagre results. If every investment were considered separately and only in terms of technological co-operation, many of them would probably appear extremely dubious. In view of this, it will obviously be an advantage for relationships to have other purposes as well. The relationships that are formed with customers and suppliers do have such an advantage: they generate continuous revenues and any collaboration that results can be regarded as an added value. Horizontal relationships do not enjoy this type of benefit and this increases the pressure to produce quick results, which in turn must make these relationships much more vulnerable. If they can be combined with some type of continuous exchange, perhaps in connection with training or similar activities, there will be better opportunities; it then becomes essentially a question of supplier or customer relations.

We can also reverse this argument by using it to underline the value of established relationships. Every relationship is in itself a resource. It provides opportunities for discovering solutions, for exploiting resources in new ways, and so on. Both the results of the relationships as reported by our companies and the expectations of these companies with respect to the relationships indicated that successful relations do lead to a series of improvements. It is this kind of unspecified outcome that makes it even more important to see the relationships as potential development opportunities because every such relationship can trigger several different lines of development. This suggests that companies should systematically look for suitable collaborative projects to conduct with partners whom they already know, rather than limit themselves to seeking partners for projects they happen to have thought of. This suggested type of search aims at exploiting investments that have already been made in relationships, and will probably be just as efficient as the apparently more rational search aimed at exploiting an idea.

Thus relationships create room for manoeuvre while also serving to mobilize resources for generating change. Existing relationships are a company-specific resource, which is difficult either to take over or to imitate. This is never more evident than when a company tries to penetrate a new area. It normally takes a long time to become established. A quicker way may be to buy a company that is already established and to use it as a springboard. Or a common intermediate way is to take over some key people who possess the personal contacts needed to launch new relationships.

Problems in relationships can also occur if a company wants to leave an area. It may be difficult to recoup investments already made, and the break-up of the relationship may be perceived as a betrayal of the other partners with a consequent loss of good will. The reaction of the partners is understandable since they may face a loss; they have generally invested as much in the relationship as the company now abandoning it. The investment aspect of these relationships can be described as something of a paradox. The company that possesses no relationships is theoretically free to enter into collaboration with anyone at all, but in fact it is difficult to find anyone who is interested. The company that has already entered into a number of relationships will find it much easier to interest a partner, but its choices will be far more limited. To acquire room for manoeuver in one dimension, the company has to sacrifice some of its scope in the other. If the company has a unique and interesting proposal of great potential it might be able to find a partner even if it has no established relationships, but we found that in general, established relationships are a vital condition for the initiation of successful collaboration. This leads us naturally to the question of the type of partner that may be interesting.

Collaborative partner

In the previous section we found that investment in relationships with customers and suppliers was easy to justify, since it probably also generates continual profit. And our analysis of results showed that four-fifths of all collaborative projects were in fact 'vertical', that is undertaken together with customers or suppliers. An interesting and somewhat unexpected discovery concerned the limited role of partners in educational institutions. Probably because we ourselves are connected with a university, we had expected this category of partner to be more important than it proved to be. Of the total 496 collaborative relationships that were analysed in greater depth, only 16 involved research institutes of any kind.

Although customers and suppliers dominate, a large group of companies makes only very limited use of partners in these categories. About half the companies collaborate with two of their ten most important customers, and an equal number collaborate with two at most of their biggest suppliers. This result suggests that there is a good deal of unexploited potential here, since we can assume that some form of relationship exists even in the other cases.

Thus for many companies it is a question of activating established relationships, rather than building up new ones. The study also indicated that nearby partners are particularly important, probably because the cost of a relationship rises rapidly with distance. The importance of proximity is particularly noticeable in horizontal relationships, but it is not altogether absent in the case of vertical relations. Several conclusions can be drawn from this. One is that all companies need to exploit their immediate environments. It is only too easy to regard the immediate neighbourhood as so familiar as to be uninteresting, while better and more exciting partners are to be found further away. But the cost and difficulty of maintaining these relationships can quickly reduce their profitability. Another conclusion is that companies should cherish their immediate environment, cultivating it in various ways so as to guarantee future collaborative openings. Here the parallel with the farmer is striking: if the farmer only bothers about this year's maximum harvest, future harvests will be jeopardized. Similarly, any reasonably large company can exploit its immediate environment in such a way as to weaken the basis for its own long-term development.

A particularly interesting result – but one that calls for further exploration before any final conclusions can be drawn – concerns the positive effect on development relationships that we found when relations were also maintained with a competitor. The existence of triangle dramas, which are a typical network phenomenon, proved at least in our limited sample to have a directly positive effect on collaboration intensity. In view of this it seems even more important to extend the analysis to include the reciprocal links between the different corporate relationships.[1] A company should consider each individual partner not in isolation, but in relation to other partners. It may well be that the best results for the individual company are achieved in a combination of different relationships.

This combinatory effect reinforces our earlier conclusion that it is more important for a company to co-operate with an old familiar partner, whose possibilities in various combinations can be easily recognized, than it is to look for a partner that will be exactly the right one for the particular problem at hand. At the same time, this result also suggests that the character of the collaborative relationship itself must also be very important.

Character of the collaborative relationships

We have already noted that relationships call for significant investment, since it takes a long time to build them up and since many people on both sides are in regular contact with one another. If we envisage every relationship as a rope, then the rope is made up of a great many thinner strands plaited together. And as the plaiting continues, both parties discover better and better ways of effecting it.

Another important characteristic of corporate collaborative relationships, which also tallies nicely with this way of describing their development, is the low level of formalization. Very few of them have any element of formal agreement, and even fewer are entirely based on such agreements. Such legal agreements as do exist generally refer to some other part of the overall relationship (annual agreements on volume or prices, for instance). The long history thus probably acts as a kind of insurance; both parties have had an opportunity to build up confidence in one another. This trust will presumably also mean that expectations on both sides are realistic, that is they are on a scale with which the other partner is able to cope. Both parties know the other's strengths and weaknesses.

Another interesting quality, which complements those described above, is that the more important relationships are generally exploited in several collaborative projects rather than just in one. It is really a question of a long, ongoing development relationship. This is evident both from the history of the relationships and from expectations about their future. The very fact that the parties have learned to co-operate with one another leads to the expectation of further collaboration.

All this suggests that the collaborative relationships generally evolve organically. They originate when a number of factors happen to coincide in an appropriate manner. It may then be interesting to ask what happens if a company starts doing more to plan its relationships? Can one in fact plan them? The answer is by no means necessarily affirmative. However, a company can obviously improve its own character as a partner in collaborative projects. Those responsible can raise their own awareness, they can learn to pick up signals from the other party, extend their interest in the other party's development, improve their handling of co-operation, and so on. There are at least a thousand different ways for a company to improve its quality as a partner, just as the handbooks claim when it comes to improving sales technique. None the less, it seems that given the character of collaborative relationships, a planning model of an over-simple or rigid kind may well do more harm than good.

Theoretical implications

In reporting these results I have discussed a great many correlations between individual variables. I have generally opened by reporting what we had expected in light of our theoretical frame of reference and rather random ideas. In several cases the correlations coincided with the direction and strength we had expected, but in many cases the results conflicted with our expectations. Our network perspective is still too incomplete for us to see with any certainty if the individual results *per se* – whether they support our expectations or not – contradict or support our theoretical model. But if we lump the results together and analyse them as a whole, it becomes possible to address this question. Moreover, the individual results will then be helpful to us in formulating a more precise and developed network model.

A first conclusion is that the three basic components from which we constructed our network model appear by and large to have been of value to the empirical study. It has been possible to identify relations between network behaviour in terms of technological cooperation on the one hand, and the characteristics, activities, and resources of the companies on the other. All three groups of variables are thus of vital importance to the way the network functions; it is possible to relate them not only logically but also empirically to the way it works. However, we know much less about the appearance of the direct and indirect relationships thus formed. Certain specific complementary approaches would be needed to give us greater insight into this aspect, and the results of the present study have suggested some that could be useful.

One approach is connected with the classification of different activities. In Chapter 2 we distinguished between exchange and production activities, a division deriving from economic theory and based on the fact that exchange activities are concerned with co-ordination and that production activities are therefore independent of one another. However, our classification has no connection at all with this theory; it is simply a relic of own research thrust 15 years ago, when we began by studying exchanges and therefore automatically distinguished these from other aspects (i.e. from production activities). In a purely network perspective and in view of the close connection we identified between customer relations and certain dimensions of the production activities, this no longer seems so appropriate. Exchange activities do undoubtedly provide links between production activities, but it can also be said that production activities provide links between exchange activities.

Thus in the network perspective the two types of activity are mutually interdependent. Theoretically, it should be possible to regard exchange as a reproduction of the production activities (and of any earlier exchanges and production activities). Series of production activities linked by exchange activities are therefore all dependent upon one another, and generally have to adapt to one another in various ways. If any one of either kind of activity is altered, changes will be required in other activities, or the input and output characteristics of the altered activity must be the same as before. Thus there does not seem to be much point in a network perspective in dividing activities between production and exchange: other distinctions or descriptive dimensions should be sought instead. One such dimension of obvious importance concerns the reciprocal relations between different activities. A fruitful starting point here might be to explore the commitment concept described in Chapter 2 and previously developed in Hammarkvist *et al.* (1982).

Another closely related approach could be to pursue the analysis of the links between the different 'triangle' relationships which I mentioned in my description of our results. The existence of positive and negative correlations between different relationships is the basis of all networks, but the triangle drama introduces yet another dimension, as can be seen in Figure 9.1.

Figure 9.1 shows three triangles of actors, consisting in all cases of two suppliers and one customer. Two are in balance (a and b), while the third is not. The first example, where everyone likes everyone, can occur when Y and Z sell complementary products which function well together to X. In this case all three companies gain from continued collaboration, and we can see clearly how the different relationships support one another. In case b, X and Y have a common enemy. For various reasons, both of them dislike Z. Naturally, their common dislike strengthens their own positive

Technological co-operation

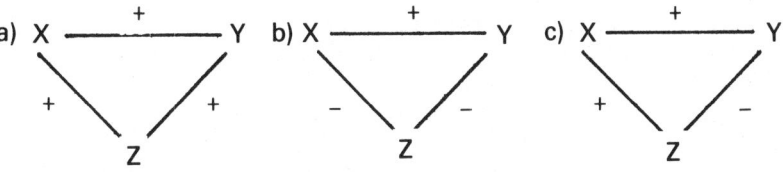

Key: X = customer   Y and Z = suppliers

Figure 9.1 Triangles: balance and imbalance

relationship. Cases a and b are both in balance, and can be expected to persist. Case c, on the other hand, is not in balance. Here X likes both Y and Z, but Y and Z dislike one another. perhaps Y and Z sell competing products, which X buys. According to all attitude models based on congruence, c would be described as being in a state of imbalance and likely to be successively adjusted towards either case a or case b. But according to our empirical results, case c produces the highest level of collaboration between X and Y (alternatively, between X and Z). The reason may be that the imbalance in itself generates greater collaboration, or it may be that even situations of so-called imbalance can persist for a long time in industrial networks with their special distribution of roles. The underlying assumption of the congruence model is that in the long run actors (i.e. people) refuse to accept the imbalance. Industrial actors (i.e. companies) may not only be compelled to accept it, but may even regard it as a natural aspect of network behaviour. Often, too, relationships are so multifaceted that they cannot be described in terms of pluses or minuses alone; rather, they are likely to include at least some of both. Moreover, the situation is often perceived differently at different levels in the company. An actor who is seen as a threat at management level may be recognized as a competent customer, supplier, or partner in the functional departments. Theoretically speaking, there is every reason to complement and extend the network model to allow for a closer analysis of these triangle dramas.

For theoretical reasons we also need to complement our categorization of the resource bases. We have distinguished five types of resources: input, marketing, capital, personnel, and technology. Our analysis of results appears to confirm the importance of all five, but there also seems good reason to add a sixth. As noted in the second edition of this chapter, the actors' established relationships also represent a resource since they extend the available opportunities. A company possessing established relationships is

Technological co-operation

able to commercialize the other basic resources more efficiently than another that does not possess such relationships. In the first five instances it is a question of the supply of the particular resource and the actor's control over it. In the sixth case it is more a question of the use of the resource, in the sense that it can be employed in different activities, provided it is combined with other resources in a specific way. Control over this network resource is certainly very important for the individual actor, and better ways of describing and analysing it must be developed.

Note

1 In a more collaborative research project with researchers in seven countries, this issue is developed further in regard to international relations. The project (IMP2) focuses on how the development in one relationship is influenced by developments in other relationships.

Chapter ten

Network analysis for corporate management

In the previous chapter we expressed some doubt as to whether the intentional and systematic planning of technological co-operation really helps to improve the quality of technological development. The alternative could be to develop a company's general capability for handling and developing relationships. It is certainly possible for companies to improve their own status as network actors, and it is in fact vital that they should do so. However, the best way of achieving this is unlikely to be by increasing central control, for instance by introducing 'activity plans'. Rather, by managing resources appropriately it should be possible to improve overall corporate capabilities. In the present chapter I shall suggest a theoretical basis for developing a working model for this purpose. The model is based chiefly on the results of the study reported in the previous chapters, but I have also noted the findings of some other studies in which I have been involved as well as experience gained from a variety of consultancy assignments. Thus I have not limited myself to the data already presented here, but have approached the issue in a broader perspective.

Our network analysis has highlighted in general terms the importance of the connections between various corporate functions, for instance between technical functions on the one hand and marketing or purchasing on the other. In more specific terms the empirical study confirmed the importance of these relationships. Technological development is an important ingredient in the processing of customers and suppliers; at the same time these partners represent a vital source of knowledge and collaborative potential in connection with technological development. Consequently, the working model described below will embrace the different functions and will address various aspects of development activities and day-to-day operations.

The model concerns the corporate resource structure as well as other questions more closely connected with action. I shall try to capture this diversity by tackling the five following closely related points:

1 network structure and corporate identity
2 corporate resource base
3 network monitoring
4 co-operation profile
5 organizing

The network structure provides the natural point of departure, together with the corporate identity within the network. This identity has been formed gradually, and it consists largely of the expectations of other related parties with regard to the behaviour of the company in question. Changes also take time, and have to be considered in relation to the overall evolution of the network. Corporate identity is also affected by the company's resource base. Similarly, the functioning of the network is closely connected with the resources that constitute its base. Thus in the first two sections below we will look at the structure of the network and the identity of the company, and then examine the significance of the resource base.

The two following sections are concerned with two areas that involve the structuring of activities: network monitoring and the co-operation profile. Monitoring embraces all the corporate activities intended to capture and comprehend changes in the network – in its way of functioning or the distribution of the network roles – and a crucial aspect here is the monitoring of technology. The co-operation profile describes how the company relates to other parties in its environment and is shaped jointly by a variety of corporate functions.

The concluding section is concerned with organization. Several aspects of the first of our four points can be linked together by the organization of corporate activities, which is therefore regarded as the most general of the network tools.

Network structure and corporate identity

A first important step is to identify the relevant network or, as is generally the case, the relevant networks. Theoretically, the identification can proceed in one of three ways. One is to start from actual interactions, and on this basis to identify what Epstein (1969) has called the 'effective network'. An empirically-based identification of this kind is relatively easy to make, and provides a satisfactory first overview. However, in an analytical and development perspective this picture is of limited value. The functional network, or what I shall call here the theoretically identified network, often gives a great deal more. As we noted in Chapter 2, a network can be identified on the basis of a particular fundamental idea or theory. The inter-actor relationships that are based on this idea, or mental model, are then identified. In reality a great many mental models overlap and blend with one another, and a purely empirical identification of this interaction produces the final result: the mix which all the mental models together have created. Changes occur as certain models evolve or command increasing response. In view of this, theoretically identified networks based on basic technological solutions, for example, are much more fruitful for an analysis of development or influences on issues than approaches based on pure empirical grounds.

Each individual actor is always included in several theoretically identifiable networks. Some of these are of minor interest to the actor, as his link with them is slight. Others are more important and the actor builds up his position by combining them in a particular way. Every actor has a more or less individual way of combining the various networks. The actor's identity consists of the role and position he occupies in the most important networks, and this in turn depends largely on the way the networks are combined with one another. The actor's identity can therefore be described in terms of the point of intersection between a number of theoretically identified networks. Thus the corporate identity is inevitably both multi-dimensional and complex. It varies according to the angle from which it is viewed. The various identities – for this is not only a perceptual phenomenon – will by no means always be in harmony; indeed, the contradictions between them can be seen as an important developmental component. Moreover, changes in the networks can also affect the individual actor's network identity; conversely, if a company (an actor) wants to change its identity in one of the networks, it can either change its position in one of the major networks or it can combine its networks in a new way. Figure 10.1 illustrates a rough checklist for the development of a corporate identity as outlined here.

The analysis consists of a comparison between the theoretically identified networks and the company, starting from a brief description of the corporate activity structure and the interaction partners.

Starting from an analysis of this kind it is possible to identify the networks which are most important to the company. Every network is described in terms of actors and activities, for example according to the basic outline provided in Figure 3.2. The next step involves a more detailed analysis of various conditions in the networks, for

Network analysis for corporate management

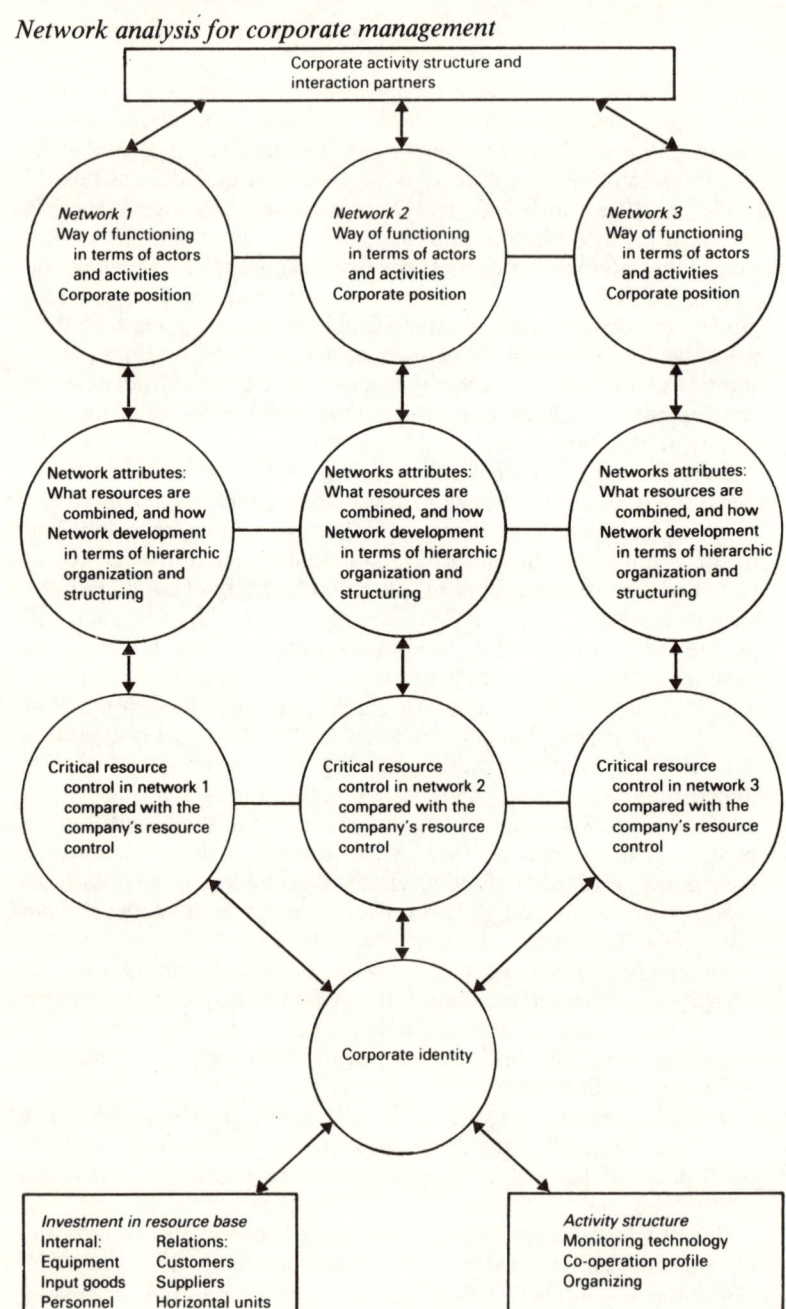

Figure 10.1 Checklist for developing a corporate identity

example what resources are combined with one another, and what development looks like in terms of structuring and hierarchic organization. An obviously important aspect is the role and function of technological development within the respective networks. The description and analysis of the various networks should make it possible at the next stage to assess the control of critical resources in the network and to consider the company's own resource control in relation to it.

Having analysed the various networks, it then becomes possible to describe the corporate identity in relation to each of them. An important point here is that corporate identity *vis-à-vis* the networks is determined partly by the company's activities, its partners, and its control of resources within the network, but also very much by the corresponding conditions in the other networks. It is also important to remember yet again that the identity need not necessarily be perceived in the same way in the different networks. The perceived identity is extremely important, since it determines the way in which other actors will react to the various activities performed by the company in question. The different perceptions are also valuable because they create a certain room for manoeuvre. Starting from a critical analysis of its own identity, the individual company will find it natural to consider changes in its investment in the resource base, perhaps, or in the structure of its activities. Corporate identity is multi-dimensional, and the question naturally arises as to whether it would be helpful to select one or a few simple dimensions such as market share. In principle, however, this is not recommended. In fact, in the long run at least, the opposite should apply, that is it is more a question of discovering new dimensions. Since the control of resources is important, the size of resources in the various areas does matter (and consequently so does market share), but this is a long way from saying that relative size is of such importance that it is always critical – any more than we can say that IQ is a predominant dimension in individual identity.

The two-way arrows in Figure 10.1 indicate that analyses are made on more than one occasion, and that changes do not only occur as the result of an analysis. On the contrary, opportunities may be discovered in many contexts, perhaps in connection with – or because of – some particular investment or in interaction with customers. Note, too, that the model includes neither economic estimates nor any calculations of market potential. Obviously, some sort of rough evaluations must be made in connection with major investments, but exact calculations are not usually of any great value in these contexts.

As an example of the way a network can be identified and

described, I shall refer to an earlier analysis of the hydraulics field in Sweden. This study was made some years ago, and several changes have since occurred in the relations between the companies concerned.

Illustration: the hydraulics network in a Swedish perspective

The point of departure for this study was a particular technology. Hydraulics is the theory of the movement of liquids in pipes and channels. It is generally regarded as the branch of technology that deals with practical applications of hydraulic engineering. Hydraulic power transmission involves the transmission of power or torque through the medium of liquid. Since it came into prominence at the beginning of the 1950s, hydraulic technology has become increasingly sophisticated and a growing number of applications have been found for it. The following technical applications can be identified:

1. hydrodynamic power transmission, used among other things for coupling in certain vehicles, generally with a small centrifugal pump, whose pressure head is caused to activate a piston connected to the regulating valve
2. hydrostatic pressing, used in brakes, for example, and involving the application of liquid pressure in a sealed space
3. the hydraulic press, used for hydraulic rams, pressing machines, hydraulic weighing machines, and so on, consisting of two cylinders, one larger than the other, linked together horizontally. A small force applied through a coupling to a piston in the smaller cylinder sets up pressure in the water, which acts over the greater area of the other piston and gives rise to magnified force.

A hydraulic system is composed of different components and subsystems, the most important of which are illustrated in Figure 10.2. The nature of the different components and the way in which they are combined means that the hydraulic system can vary in character, that is they can be adapted to different applications.

I shall now briefly describe the different components and application areas.

Component areas

Engines and pumps. These constitute the power core of the hydraulic system and are produced in large batches by a few big manufacturers. Most of these are large international companies,

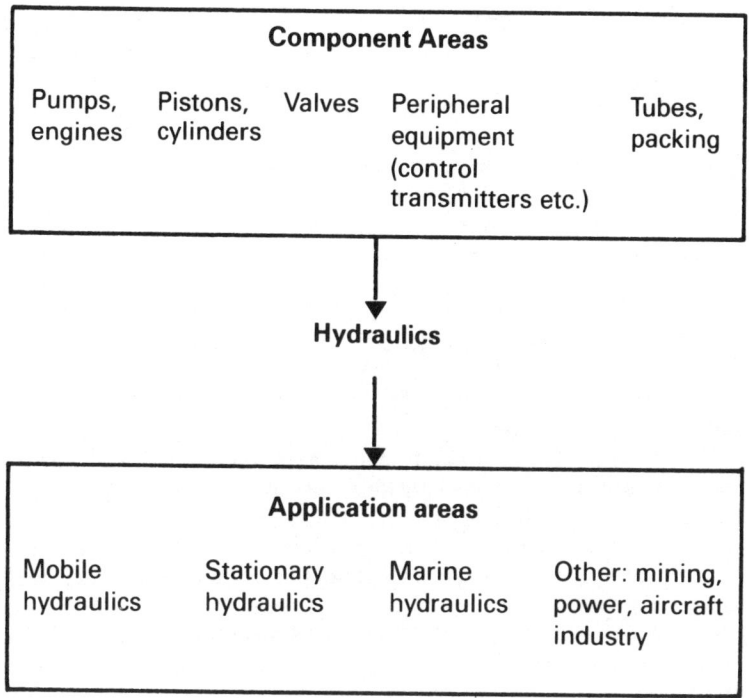

Figure 10.2 Theoretical structure of the hydraulics network

whose hydraulic components represent one of several product lines. The level of standardization is comparatively high, and customers choose from an established range. The area is technologically stable and changes are small, occurring within the framework of the established network. There is very little likelihood of major dramatic changes in the short run.

As regards both products and structure, this area has found its form. Competitive advantage is achieved by streamlining production and distribution and exploiting the advantages of scale.

Swedish engine manufacturers are Volvo Flygmotor and Hägglunds, which also manufacture engines in the United States. Products are divided between the companies, so that Hägglunds concentrates mainly on large engines. The Swedish manufacturers of pumps are Volvo Flygmotor and Sundins. There are substantial imports of both pumps and engines. Important companies are Bosch, Rex Rotary, Sperry Vickers, and Kracht.

Pistons and cylinders. Some changes can be discerned in this product area. At the manufacturing level the structure is by no means as concentrated as in the case of engines and pumps. The manufacturers are predominantly Swedish, and include several small or medium-sized companies. Production is determined to a fair extent by customer orders, and the manufacturer's status *vis-à-vis* the customer resembles that of a subcontractor. In several cases the manufacture of pistons and cylinders is an integral part of a company's hydraulic operations.

Cylinders are often made of light metal or steel. Roughly speaking, in the manufacture of cylinders a tube is cut and supplied with ends. It is extremely important that the material in the tube is as free as possible from airholes, and that the thickness of the metal is constant. Both conditions affect the quality of the waterproofing between cylinders and pistons.

At present there is a good deal of experimentation with various new materials such as carbon fibres and plastic sheathing. There is also talk of developing an international measurement standard. Technological advances in the production of steel tubing have simplified the manufacture of cylinders. To achieve an acceptable level of proofing between piston and cylinder, the tube has to be completely cylindrical and its inner wall absolutely smooth. In this respect the so-called DOM tube has excellent qualities. This tube is manufactured and marketed in Sweden by Virsbo Bruk and SKF. There is also a certain amount of import. The DOM tube was originally an American innovation. In the course of adapting the tube to hydraulic use in Sweden, technological co-operation was established between the manufacturer of the tube (Virsbo Bruk), the manufacturer of the waterproofing (SKEGA), and the hydraulics manufacturer (Hiab-Foco).

Valves. The valve is the heart of the hydraulic system. Its function is to regulate the pressure of the liquid in the tube coupled to the valve. This consists of a cast valve-box, in which the pressure is regulated by inserting the pistons in a particular way. The valve is a sensitive part of the piston, and the competitive potential of a complete hydraulic system is largely determined by the quality of the valve. The manufacture of valves is a technologically skilled operation with high precision requirements. Certain parts of the manufacturing process originated in the car industry, in the manufacture of gearboxes. Swedish companies do make valves for a few areas of application, but other applications require a high level of technology that is not at present available in Sweden.

The valve is an extremely crucial component that has to be strictly controlled. The character of the hydraulics network is largely determined by the technological design and control of this component. Changes here would have dramatic effects on companies operating in the hydraulics field. The main valve manufacturers in Sweden are Monsun-Tison, Hiab-Foco and Nordhydraulik. When the requirements are exceptionally high, as they are for stationary applications for instance, valves are generally imported, often from the German hydraulics company Kracht GmbH which is owned by Bahco.

The valve is regulated by peripheral equipment (control devices, transmitters, etc.) of various kinds. These components or subsystems transfer control impulses to the valves, which in turn regulate the pressure of the liquid. This particular component area is currently undergoing a period of rapid technological development. By making greater use of electronics, higher levels of precision and accuracy are being achieved. Technical changes in this field can sometimes have a dramatic effect on the competitive situation by radically affecting the final product (i.e. the work environment of excavating machinery operators).

Finally, we have the heterogeneous component area of tubes, proofing, and nipples. Here we find large companies manufacturing tubes and proofing, as well as much smaller companies in the field of nipples and couplings. There is a certain amount of development in tubing on the reinforcement side. Most tubing is steel-reinforced, but some attempts are being made with carbon fibres. Packing and O-ring gaskets are currently a fairly stable area in terms of current technology. The competition is international and the products are standardized.

Tubes are manufactured by the Trelleborg Group. Most of the hydraulic tubing manufacture is located in a subsidiary in Holland. Packing is made by two companies in Sweden. One is STEFA, which is a Trelleborg subsidiary, and the other is Skega in Ersmark outside Skellefteå. Imports are substantial and dominated by the Simrit Company.

Generally speaking, tubing, packing, and nipples are sold directly to the hydraulics manufacturing companies, with a secondary market for replacing the products as they wear out. This secondary market consists of wholesalers or retailers who sell the products to small local tubing service companies. These are not generally the same companies that dominate on the two submarkets.

By and large competitive strength in the hydraulics field depends on certain qualities of the valves and the peripheral equipment, and to some extent also of the cylinders. In all areas of application, these classes of components are part of the hydraulic system. However, applications are often so specialized that certain companies concentrate on a particular area – mobile hydraulics, stationary hydraulics, and so on. I shall now briefly describe the nature of the different areas of application.

Areas of application

Mobile hydraulics. This equipment is used on lorries, excavating machines, refuse lorries, fork trucks, forest machines, loading machinery, and so on. This segment is characterized by a relatively low level of technical complexity, and its market by high priced-related competition. There does not appear to be much potential for growth, and volumes vary with general levels of economic activity. Swedish companies command an impressive global market share. Batch production prevails, and a given technical design is made competitive by efficient production and marketing.

Hydraulic technology was used at a very early stage in the mobile field. The technology is thus established and fully developed. Only some very dramatic technological change could affect the situation in this area. Technological development is mainly limited to control and regulatory equipment.

In this field certain national specialities have evolved, partly depending on the way other systems are designed. In the United States, for instance, truck cranes are used less often than in Sweden, and loading and unloading is done by stationary hoisting gear instead. Hydraulic transmission is also more common in excavating machines in Europe than in the United States. In mobile hydraulics, pressure is employed within the interval 100–250 bars, which is reckoned as the average pressure.

Stationary systems. These are fixed systems installed in manufacturing industry, engineering plants, and so on. In this category we include applications of varying weight, and with working pressure varying from less than 100 to over 1,000 bars. Here growth potential is regarded as good, and the area is developing fast. New applications keep appearing, and components and subsystems are being developed. In the case of components the advances are mainly in control and regulatory equipment, for example equipment for controlling the movement of pistons and valves.

In production equipment and applications in products (OEM customers), development is conducted by hydraulics manufacturers and customers in collaboration. The supplier often acts as technical problem-solver, with the result that systems are tailor-made for certain customers. The market is more nationally orientated than the market for mobile hydraulics.

Marine hydraulics. This field is concerned with equipment for seagoing vessels and offshore operations. It is normally a case of medium-high pressure, between 100–250 bars. Mobile and stationary equipment is used. There are good prospects for growth in the marine field. Quality requirements are very high, because of the susceptibility to corrosion in aggressive environments. However, the hydraulics *per se* do not differ from other applications.

In the marine hydraulics field some companies have specialized in complete marine hydraulics control systems and the operation of large valves on tankers, offshore platforms, and others. Companies specializing in the marine sector are often established in Germany, Holland, and England. Altogether the marine hydraulics field is characterized by rapid change in both technology and structure.

Other hydraulic systems. Other systems are used in the mining, power, and aircraft industries, often requiring high pressures of 250 bars or more. In some sectors there are good growth prospects: in the mining industry, for example, there is a renewed interest in coal-mining. The various fields have their own special requirements, for instance in mines where there is a risk of fire or explosion, while the aircraft industry includes highly advanced technical applications.

In conclusion it is worth noting that the two most interesting areas of application, mobile and stationary hydraulic fields, are in some respects essentially different. The first is associated with a stable and well-constructed network, in which the roles of the different companies are clearly defined. To some extent the opposite is true of stationary hydraulics, a field that is undergoing a process of change. Here roles are not defined, and the network is much less structured. Companies compete actively with one another in various ways, and there is far less openness towards external researchers, for example, than there is in the mobile field.

Thus we have seen that the different networks that can be identified in the hydraulics fields, for instance according to components or areas of application, vary greatly in their actors, their activities, and their resources. Some actors try to operate in various network combinations connected with the hydraulics field, something which is also reflected in their identity, as well as their resource base and the structure of their activities. Other companies instead combine a network in hydraulics with a network in some other technological area such as electronics or rubber. This brief survey again demonstrates the unique characteristics of every corporate identity, although we have not yet touched upon the special character of each company's customer and supplier relationships.

The company's resource base

The second aspect that we identified earlier is the company's resource base, which was also an important component in the analysis of the corporate identity in the previous section. Making changes in the resource base is one of the ways in which a company can alter its identity. Technological development activities are also directly linked to the resource base, since they can influence it and be influenced by it.

In an analysis of the resource base an actor can first look at the current state of affairs – that is, what control the company enjoys over resources – and then try to find out how the situation evolved: in other words, to identify the resource mix which is important to the company's long-term standing. From time to time the emphasis can shift dramatically, as seen in Figure 10.3 which describes the resource base of a steel company over a 75-year period. Thus resources of the future are often likely to be different from those controlled today, and yet corporate decision-makers generally pay more attention to the situation of yesterday. Even well into the 1970s, some Swedish steel companies were still making investments on the raw material side (e.g. in the mines), although the importance of this particular resource was dwindling. The point of the second part of our analysis – the retrospective analysis – is to identify major development trends. Technological development is obviously a major factor since it can help to reduce the importance of certain resources, perhaps by making it possible to replace one input item or one type of professional capability by another, and it can also link resources together in new ways.

The next stage in the analysis concerns relations with various partners such as customers, suppliers, research institutes, and so on, and how they link up with the resource situation. How can co-operation on technological issues be utilized to extend relationships to new partners or to improve the quality of existing relations? Conversely, actors should try to find out whether any of their partners are developing new resources over which they would like to acquire some control themselves. Collaborative technological projects can provide a suitable means of acquiring knowledge and experimenting with new resources.

All this may call for a build-up of

- technical resources (investment in machines and plant)
- knowledge base (new technological areas and applications, people, patents, etc.)
- social networks (close contact with key people among customers, suppliers, research institutes, etc.)

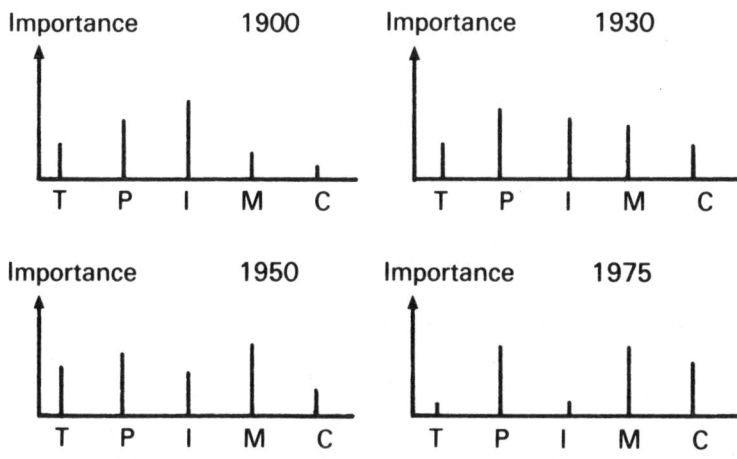

Source: Axelsson and Håkansson (1979).

Key:
T technology
P personnel
I input goods
M marketing
C financial capital (for definitions of the different types of resources see Table 2.1)

Figure 10.3 Resource base of a steel company over a 75-year period

Technological development, including that carried out in collaboration with a variety of partners, represents an integral part of this resource build-up.

Network monitoring

Changes that affect the individual company can occur anywhere in the networks to which it belongs. Development assumes so many forms, occurs in so many places, and is surrounded by so many interdependent relationships that no complete monitoring system is ever possible. A variety of flexible monitoring arrangements will always be necessary. In other words, there is no possibility of creating an effective and integrated management information system. However, the matter should not be neglected just because there are difficulties because the company will not automatically hear about changes that may be important to it in the future. I shall focus here on two aspects: first, where the important technological changes occur and in what form; and second, existing or desirable channels for the transmission of knowledge.

Even if we limit ourselves to technological development, the interesting locations for change and the forms it can take are legion. However, the two types of technological development which we have identified earlier – leap-wise and step-by-step – can provide us with a point of departure. We could also classify changes as affecting products or production. By combining these two dimensions with the main types of partner, we obtain a survey matrix that could be useful to an analysis of the development situation. The general form for such a matrix is illustrated in Figure 10.4.

	Step-by-step changes		Leap-wise changes	
	Production	Product	Production	Product
Own company				
Customer A, B, etc.				
Suppliers X, Y, etc.				
Customers' customers a, b, c, etc.				
Distributors				
Parallel units A, B, etc.				
Competing units A, B, etc.				
Other units a, b, c, etc.				

Figure 10.4 Matrix for monitoring technological development in corporate networks

The matrix can be used to steer data collection, and it can provide a model for identifying the combinatory effects of disparate changes. In this way the company should obtain first an overall picture of the leap-wise and step-by-step technological changes known or expected to be taking place in different parts of the network, and second, some idea of the areas in which corporate information is inadequate. The latter will suggest what further investigations are needed, based on an analysis of the threats and opportunities created by the changes. This in turn will provide a basis for actions such as mobilizing forces against the threats or exploiting the opportunities in various ways; it will also of course open the way for corporate initiatives.

The model illustrated in Figure 10.4 might give a false idea that it is fairly easy to get a satisfactory picture of the development situation. This idea fades away immediately when we begin to look at what is required of the information channels, if these are to be at all adequate.

Network analysis for corporate management

We can start from an earlier study of the transmission of knowledge between the research and industrial worlds (Håkansson, Laage-Hellman, and Axelsson 1983). Figure 10.5 provides a summary of the situation, classifying the channels as 'one-way' or 'interactive'. The one-way channels are then subdivided into 'direct' and 'indirect'. These classifications are based on the expected effect of the knowledge transfer over the various channels. It is assumed that knowledge transmitted over the interactive channel is more apt to lead to direct applications in industry. We can also put this the other way round and say that if knowledge is to be transformed into practical action (i.e. applied), interactive channels are necessary.

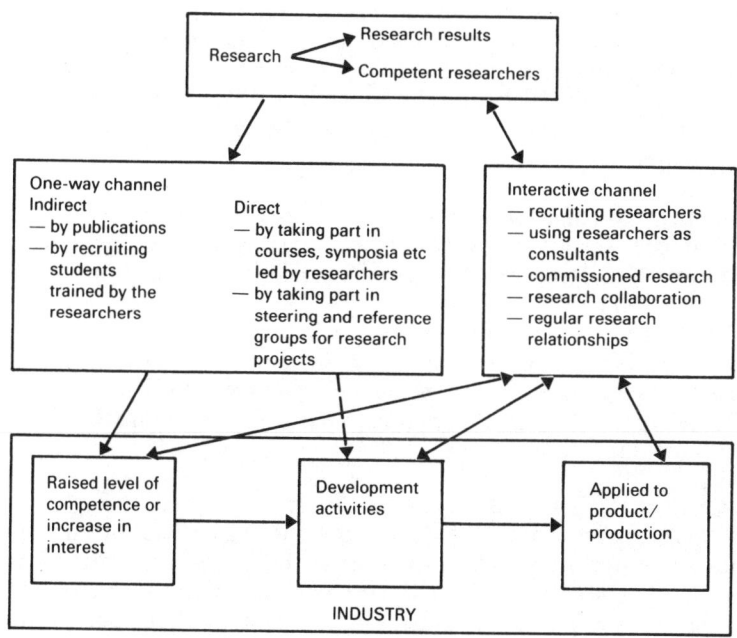

Source: Håkansson, Laage-Hellman, and Axelsson (1983).

Figure 10.5 Knowledge transmission channels and their effects

One-way channels can transmit knowledge that is useful for general purposes of orientation or competence-raising. Much more rarely, however, does knowledge transmitted over these channels lead to a direct change in operations.

145

The difference between one-way and interactive channels depends on the fact that if the two sides (research and industry) are to influence each other more directly, certain conditions must be fulfilled which can only be met by the interactive channels. We can define these conditions as a common language, mutual knowledge of one another, and mutual confidence in one another.

Common language. The most important and also the most difficult problem is probably language. Knowledge development almost always involves language development as well in some form or other. Unfortunately, however, the reverse does not apply: language development does not always mean that knowledge is being generated or enhanced, and technical or professional jargon often conceals a deplorable lack of knowledge. But, to return to the first problem, it has to be acknowledged that learning new languages is a laborious business. It is not enough to listen to them; we have to learn to work with them ourselves, and in so doing often to develop them further. This in turn requires an investment of time and motivation.

Mutual knowledge of one another. If an effective link is to be maintained, the units involved must know something about one another. For instance, they need to know something of their partner's competence and resources. They also need to know how the other party's organization works, whom to contact on what questions, and so on. Accumulating this kind of knowledge takes time and puts pressure on personnel resources. It is mainly by interacting with one another that both partners can acquire this kind of knowledge.

Mutual trust. There is always some risk involved when new knowledge is being used; the result can never be guaranteed from the start and unforeseen problems may arise. Taking risks like this calls for mutual trust. Both sides must feel confident that the others are serious and are putting their cards on the table. Without a stock of trust the relationship is likely to be short-lived. Mutual trust is also necessary if the exchange of information and knowledge is to be as free and open as it should be. Mutual trust generally builds up gradually as the two parties first test one another in situations of minor risk (meetings, lesser projects, etc.), and then proceed to more important and more difficult cases.

Figure 10.5 is concerned exclusively with the transfer of knowledge between the worlds of research and industry, whereas here we are also interested in the transfer of knowledge between industrial companies. Theoretically, however, the same arguments must apply. The difference is that in our present perspective research and researchers are not the only interesting segment; technicians and technological development in general are equally crucial.

For the monitoring of technology, which is only one part of network monitoring as a whole, the company needs a variety of channels to several different partners. And the greater the need for quality in the knowledge transfer, the more costly will be the necessary channels. Thus many choices will have to be made, for instance about the quantity or quality of the actors. The knowledge network must be continually reviewed, developed, or adapted to changes in the company itself or in its environment.

Co-operation profile

Co-operation in a variety of forms has been part of all the points we have discussed so far. It represents one way of building up the resource base, and it is of crucial importance in developing the corporate identity. Also, as we have just seen, it provides an important channel for the transfer of knowledge. In other words, there are countless reasons for co-operating and equally many forms that such co-operation can assume. All this variety could be used as an argument against discussing or trying to control the overall co-operation profile of a company. Such attempts might upset or reduce the variety and thus do more harm than good. However, the great variety does not arise spontaneously, and in some companies – as our study has shown – the number of co-operative undertakings is extremely small. In other companies the reverse is true, that is there are so many opportunities for possible co-operation that it is necessary to choose between them. Taken all in all it thus seems that all companies would benefit at least now and then from examining their co-operation profile to see just what is going on. They can examine, for instance, whether a reasonable proportion of total development activity is being conducted in external collaborative projects, whether the various kinds of collaborative partner are appropriately exploited, and if existing collaborative relationships are being conducted efficiently. Let us now examine these questions.

In a previous section we discussed the correlation between the proportion of external collaboration and profit or growth. According to classical marginal analysis, we would expect on average a reverse U-shaped relationship, that is the best solutions in terms of profit and growth occur when the company invests about half its development volume in external collaboration. There are obvious economic reasons for maintaining a certain amount of external collaboration, and there are equally good grounds for not having too much. If the company is to be able to exploit its external collaboration, it needs a certain level of internal activity as well to hold things together. Without this, the various collaborative projects can easily become fragmented. The empirical findings did not contradict these hypotheses, but in the case of the profit dimension indicated considerable variation. What constitutes an appropriate proportion of external co-operation thus varies in relation to the company's situation, which in turn is probably related to both its stage in the life cycle and the network to which it belongs.

A small company with one or a few large customers, for example, may tie up all its technological development with these same companies. On the other hand, if a company has a strong technological standing in its own field, it can acquire competence and knowledge without embarking on collaborative projects. In such a situation an introverted strategy may be perfectly viable. Furthermore, it seems quite likely that companies may need to withdraw somewhat into themselves at certain times, while at others they are greatly in need of association with diverse partners. Similarly, changes in the company or in its networks can motivate consideration of a change in the external share.

The second question asked whether various types of collaborative partners were being appropriately exploited. The main difficulty here is that it is seldom possible to make a completely free choice of partners, since the special interests of the partner also come into the picture. It may often be better to choose partners who are interested and motivated, even if they may not be the best in all other respects, rather than trying to motivate partners who do possess the best qualities otherwise.

This was in fact very much the picture that emerged from our empirical study: companies collaborate with partners who have shown that they are interested. The distribution of co-operative projects among forward, backward, or horizontal relationships suggests this. The absence of collaborative partners of one particular kind can probably be partly explained in these terms, but it is also obviously a weakness for the individual company. Many companies are missing opportunities which would not in fact be very costly to exploit.

The third question asked whether existing collaborative relationships are being exploited to the full. Our study provided no answer here, except in the personal impressions we formed during the interviews. Some of the companies appeared to be well aware of the importance of their collaborative relationships, and probably therefore exploited them extensively. But these companies were few. Most revealed no such awareness, and there is certainly a great deal of under-exploited potential here.

The answer to the question of whether the collaborative relationships are being conducted efficiently is probably similar. There is certainly plenty of room for finding more efficient forms of collaboration. The cost of every relationship is considerable, as we could see from the contact structure of the companies studied. Moreover, we found very few companies that had made any systematic attempts to improve the outcome of a relationship by reducing its costs, for instance, without also spoiling the positive effects.

Altogether we concluded that the great majority of companies would benefit from a more systematic processing of their overall co-operation profile and of their individual relationships. There is a considerable potential for improvement in this direction.

Organizing

The relationships we have been studying represent complex processes that persist over long periods of time, often including many people on both sides. The partner concerned may be a customer, a supplier, or what we have designated a horizontal unit. From various detailed case analyses we have also found that each link in itself – at least each fairly considerable link – can be further divided into subrelations which are only partly dependent upon one another.[1] Furthermore, there are various interdependent connections between the various relationships. Triangle dramas are common and significant, and as we have seen there are also connections between the individual company's relationships and changes within the network. Each relationship consists of a number of activities of a technological, commercial, and social kind. Altogether there are a great many actors (people, parts of companies, companies) performing a great many activities, which are dependent in various ways upon one another. All this calls for a lot of organizing.

This organizing activity has at least three major functions. The first is to co-ordinate activities in order to enhance their efficiency. For example, a company's relationships with its various suppliers must be co-ordinated with one another and with the company's own production activities, as well as with developments in various customer relationships. Thus co-ordination has to cross the functional borders, which can cause difficulties, particularly in large companies.

The second aspect concerns the role of organization in activating and motivating diverse actors to some common action. Organizing in this context can be called 'networking', and it implies a more or less systematic attempt to threaten, attract, or persuade actors (again, these may be individuals, parts of companies, whole companies, or other organizations) to accept or support various types of change. Since every actor is unique, the persuasive approach has to be individualized. Every actor who wants to cause change must seek all kinds of alliances, even including the 'unholy' kind in which the only shared interest may be a common enemy. Once again, the organizing activity has an internal and an external dimension, as the home organization has to be integrated in various ways with several partners.

Organizing is important in another respect. Keeping experience alive, seeing that it is spread not only inside the organization but also within the relevant networks, is an important condition of further development. Experience is not automatically accepted. And yet we know that much new development emerges from the combination of different types of experience. For both reasons the nourishing and dissemination of experience has to be organized, not usually by mechanical memories but above all by the people who have been part of it. Recruiting from and 'exporting' personnel to important partners, staff rotation, and so on are all important elements in this organizing process.

The predominant organizational solution in the business world at present is to divide operations among independent units, which perhaps function better in these respects than the traditional functional organization. Purchasing, production, and sales are brought closer to one another, which also makes it easier to link supplier, customer, and horizontal relationships together. One drawback of this greater independence, however, is that new combinations of resources or activities can easily be overlooked. Moreover, there may be such a strong emphasis on short-run profit goals that technological development – which calls for long-term collaborative activities together with many partners in order to mobilize support for development – becomes much more difficult to achieve.

New ways of organizing, perhaps in the form of networks, will probably be needed if the kind of dependent relationships described here are to be adequately managed.[2]

Notes

1 One example is Liljegren (1988) who examines the development of seller-customer relationships over a period of 10 years between Atlas Copco (a

Network analysis for corporate management

manufacturer of mining equipment) and the construction company ABV.
2 There is certainy much to learn in these matters from Japanese companies as shown in, for example, Yoshino and Lifson (1986).

Chapter eleven

Network analysis for industrial policy

Industrial policy is generally concerned with all kinds of development problems: how a whole counry is to be developed, perhaps, or a particular industry or sector, or what can be done to make a region more attractive to industry. But the problems may also be more specific than this, a question perhaps of individual development projects which are not directly profitable in themselves, but whose consequences for certain other actors none the less make them worthwhile. Or again, some organization – a research unit, a basic industry – may need support so that it can fulfil an important role in complementing other actors. Development problems may concern individual companies or the infrastructure that supports them in various ways.

All these types of problem can be analysed to good effect in network terms. Existing networks may either support or obstruct the desired changes, and so it may be a question of identifying the opportunities offered by the networks, or of investigating ways of resolving any obstructive commitments there. This last is very important because networks can represent powerful barriers to changes that in societal terms are thought to be necessary. We have already noted that the development process in networks is political in the sense that the actors involved try to steer events in directions to suit themselves. Networks know nothing of the kind of self-regulating forces that appear in the theoretical market model. Consequently, ideas and proposed changes are evaluated by the actors not in terms of their intrinsic value but in terms of how they can be used and, accordingly, how they can affect different networks.

Thus the network analysis is just as interesting and necessary in any consideration of industrial policy as was previously claimed to be the case in corporate analyses. Indeed it has been used in a number of consultancy projects connected with industrial policy issues. To illustrate the usefulness of the approach I shall briefly describe three projects below, looking at their goals, the analytical method adopted,

Network analysis for industrial policy

and their results. In a concluding section I shall summarize the chief advantages of the network approach. The first project concerned the analysis of research activities in a whole network, and their relationships with one another. In the second project we analysed the industrial connections of a number of major university research projects of a technological nature. The third concerned a regional analysis, looking at a broad technological field and trying to discover what the regional authorities could do to complement and stimulate it.

Biotechnological research in Sweden: research volume, research focus, and collaborative patterns

STU, the Swedish National Board for Technical Development, is a government agency for initiating and supporting technical research and development (R&D) work of strategic importance to Sweden and the Swedish economy. It does this mainly by providing financial and consultative support to public research institutes and private industry.

As in many other industrial countries, biotechnology has been identified in Sweden as a field with a future, and one that should therefore be given priority in the allocation of state research grants. One result of this attitude was the establishment in 1985 of a national committee for biotechnology. It is located at STU and includes representatives of industry, the universities, and the grant-allocating authorities. Its purpose is to promote a Swedish research strategy in the biotechnological field, and to stimulate co-operation and knowledge transfer between companies and research institutes.

To provide some background for its long-term development of a national strategy, STU initiated a survey of biotechnological R&D activities in Sweden. The study (Laage-Hellman and Axelsson 1986) describes the situation in 1985 in both quantitative and qualitative terms, and looks at developments over the previous 10–15 years. It presents volume data and describes the R&D activities of the various units in terms of operational focus, historical development, strategies, organization, and – not least – the frequency of external collaborations and contacts.

Data were collected through interviews with people in 33 'units', that is 18 research institutes and 15 industrial companies. For each unit a case description was produced and respondents were asked to comment and to make any additions they thought necessary.

Many types of actor belong to the biotechnological network. However, it is possible to identify a few major categories which, together with their links with one another, are illustrated in Figure

Network analysis for industrial policy

11.1. At the top of the Figure we find the pure research institutions: university institutes, industrial research institutes, and the state research bodies (e.g. the National Bacteriological Laboratory). There is no homogeneous industry that can be called 'the biotechnological industry'. Biotechnology can be defined as a field of knowledge and technology with applications in several industries and sectors of society. But if we are to speak of a biotechnological industry as such, we could perhaps best define it in terms of the companies that use biotechnological processes and methods in production – companies in the pharmaceutical, chemical, and food industries, for example. The manufacturers of certain types of diagnostic equipment could also be included. The users of biotechnologically manufactured products include final consumers (e.g. food consumers, hospitals, and clinical laboratories), as well as other industrial companies that process biotechnological semi-manufactures to produce final products.

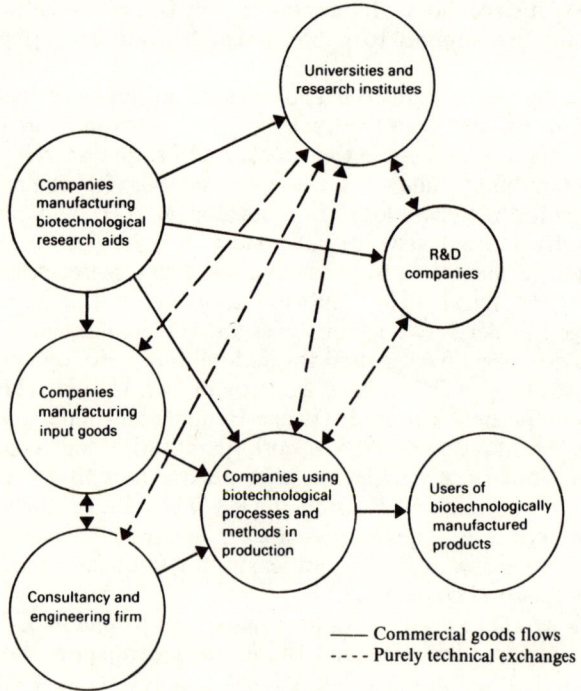

Source: Laage-Hellman and Axelsson (1986).

Figure 11.1 Type of units included in the biotechnical network and the most common relations between them

Companies manufacturing input goods for biotechnological production also belong to the biotechnological network. This group includes processing equipment (e.g. fermenting and separating equipment) as well as chemicals, in particular various separation media. Chemicals of this kind are often manufactured by biotechnological methods. As well as the regular equipment suppliers, there are also a number of consultancy and engineering firms specializing in biotechnology.

Another group of actors included in the network consists of companies making biotechnological research aids, that is special products that are used in biotechnological research. Here we find, for example, the manufacturers of apparatuses and chemicals for biochemical separation and purification, and the producers of various 'tools' (e.g. analytical reagents and enzymes used by molecular biologists in research institutes and industrial laboratories).

By R&D companies we mean companies whose activities at present consist largely of research and development. It may be a case of independent research companies founded by an established industry, or of offshoot companies started by academic researchers with or without the collaboration of external finance and/or entrepreneurs. Companies of this kind often occupy a kind of intermediate position between basic academic research and established industry.

It should be noted that Figure 11.1 does not provide a picture of the actual appearance of the network; it simply indicates the theoretical design. In reality the biotechnological network is extremely complex. It consists of a great many individual units, each with an abundance of more or less well-developed links with other units. Let us look for instance at the genetic engineering network, which is just one part of the overall biotechnological network and includes units directly involved in the development and application of genetic engineering techniques (e.g. rDNA technology).

Figure 11.2 provides a rough outline of the appearance of this network in 1985 when we were making our study. Here we have indicated the units that were regarded as crucial in industrial application, and some of the more important domestic and foreign collaborative arrangements and ownership conditions. Figure 11.2 shows that KabiGen occupied in some ways a central position in the network; for example, the company collaborated with a great many Swedish research institutes. KabiGen's central position is easy to understand, since at the time it was the only major Swedish R&D company engaged in genetic engineering.

Outside KabiGen there was little genetic engineering research in Swedish industry, and companies relied largely on external

Network analysis for industrial policy

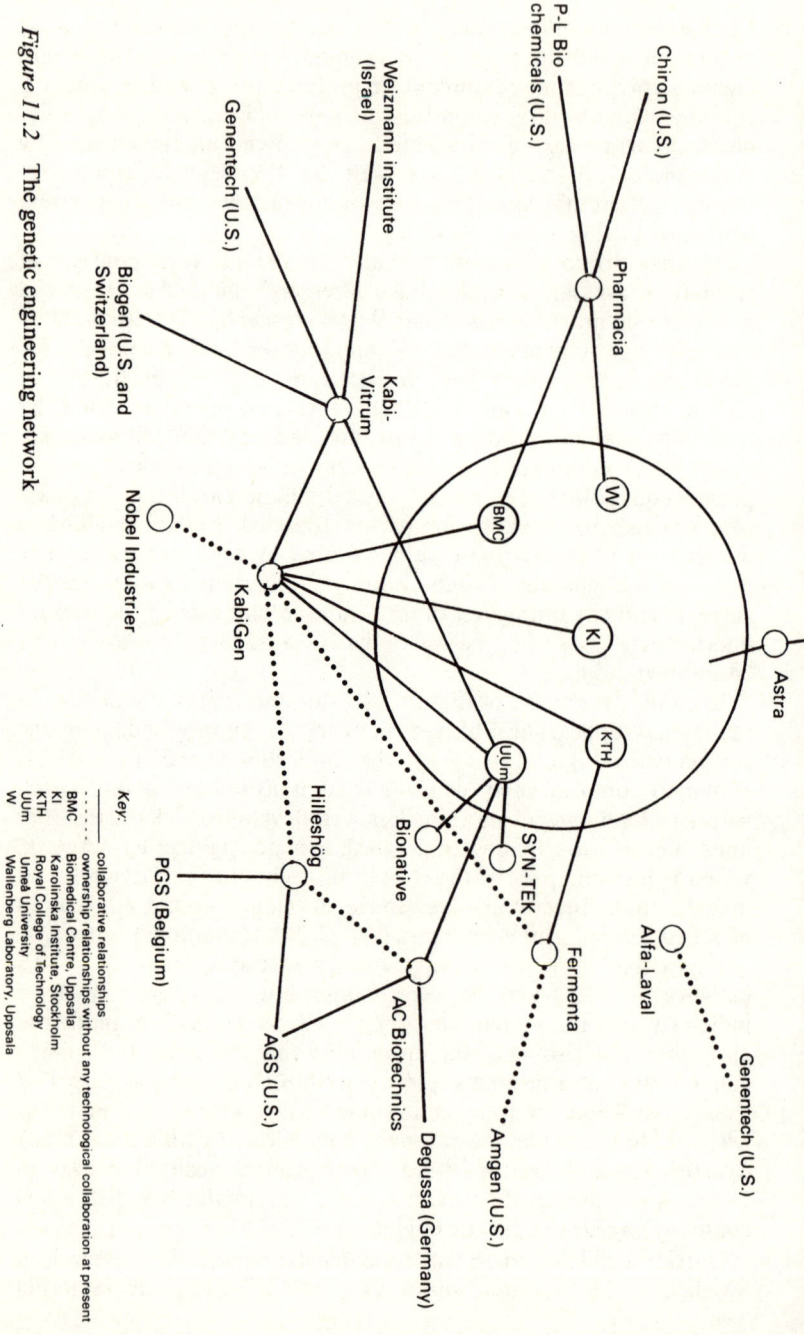

Figure 11.2 The genetic engineering network

resources. KabiGen was involved in several genetic engineering product-development projects at KabiVitrum, but this was an exception and the company has not otherwise had much visible impact on Swedish industry. Typically, Swedish industrial companies would ally themselves with foreign – generally American – genetic engineering companies. Thus KabiVitrum had projects at Genentech, and at Biogen's two laboratories in the United States and Switzerland, despite being part-owner of KabiGen. One reason, of course, was that the foreign R&D companies were thought to be more advanced than their Swedish counterparts. Another collaborative partner was the large Weizmann Institute in Israel.

Pharmacia had started fairly recently to develop a molecular biological research programme of its own. The idea was to acquire a level of competence that would enable it to act as a qualified buyer. The chosen strategy was to invest in external partners rather than to build up genetic engineering research on a large scale within the company. The idea was to try to utilize the innovative biotechnological companies that were springing up everywhere, particularly in the United States. In January 1986, for example, a collaborative agreement was made with the American genetic engineering company Chiron Corporation. Another partner in the United States was P-L Biochemicals, which Pharmacia had bought a few years before. The company specializes in products used in molecular biological research.

Hilleshög was one of the Swedish pioneers in genetic engineering. Since it is interested in applications in the plant-breeding field, it acquired a share in KabiGen some time ago. But as KabiGen was unable to invest in plant breeding, Hilleshög decided to look overseas instead. Together with foreign partners Hilleshög has started up two genetic engineering companies, Advanced Genetic Sciences (AGS) in the United States and Plant Genetic System (PGS) in Belgium.

There was thus a clear tendency for the Swedish biotechnological companies to import genetic technology, concentrating their operations instead on processing technology, that is learning to handle the genetically modified micro-organisms/cells at the cultivation and development stage, and on the clinical aspects. KabiVitrum was and still is the only Swedish company with a product on the market (the Chumanc growth hormone, Somatonrom, which was approved for marketing in 1985). Alongside the development of new products, the company was also working on improvements in existing manufacturing processes on the fermentation and purification side. It has also produced a growth hormone with the exact amino acid sequence (191, instead of 192 as in its original product). Here there has been a certain amount of co-operation with Genentech, which has the sales rights in the United States and Canada.

The collaboration between Pharmacia and Chiron is concerned primarily with the development of drugs based on a biosynthetically manufactured enzyme known as hSOD (human super oxide dismutase). A joint venture company has been established for this purpose, to be located at Chiron's plant in California. It will conduct pre-clinical and clinical trials and prepare for registration and introduction on the market. According to the collaboration agreement, Chiron will contribute knowledge of genetic engineering and related processes, and will be responsible for cultivating the original substance on a large scale with the help of genetically modified yeast cells. Pharmacia is to supply data from its earlier studies of hSOD and will be responsible for the final purification and chemical processing, and will help with the registration routines. The final products will be marketed by Pharmacia. This division of labour was expected to utilize the capabilities of the two companies to optimum effect.

Thus, apart from KabiVitrum and Pharmacia, things looked pretty thin on the product development side. The commercial applications of current research were generally still in the remote future. Of the various research institutes, those of immediate interest to industry were the application-minded units at the Royal College of Technology in Stockholm and at Umeå University. Some other units were in the process of extending their industrial contacts, however, and this was regarded as a positive sign on the industrial application front.

In the quantitative part of the study it was found that there had been a substantial increase in the volume of R&D investment in industry and in the public research system during the first half of the 1980s. In absolute terms the increase had been greater in industry than in the universities and research institutes, with the result that qualified researchers, especially in the fields of cell and molecular biology, were in short supply. This was one reason why companies had been more inclined to locate their genetic engineering research abroad. But it should be added that the trend is still much the same now (1988) as that described by us in 1985, and the tendency for Swedish companies to conduct genetic engineering research abroad is still strong. However, the way this is done can vary. During the past three years Sweden's largest pharmaceutical company, Astra, has established a biotechnological research unit of its own in India. This is associated with the well-regarded Indian Institute of Science in Bangalore, with which Astra had long enjoyed good relations. Pharmacia has established a company of its own in California to conduct genetic engineering research. And several other Swedish industrial companies have bought shares in American genetic engineering companies (among others Aritmos, Karlshamn, and Perstorp). California Biotechnology has also been involved in the establishment of a new biotechnological research company at Huddinge Hospital in Sweden.

The overseas location of so much corporate research has had positive and negative effects on the development of biotechnology in Sweden. On the positive side, companies have been able to extend their resource base and their competence, and to compensate for imbalances in the supply and demand for resources at home. At the same time it indicates that the Swedish university system is generating too few researchers (and the wrong type of researcher, according to some people in the industry), and perhaps actually serves to aggravate this problem. Because of the lack of qualified research personnel and the relatively low biotechnological capability in much of Swedish industry, there is a risk that Sweden may fall behind in the commercial exploitation of modern biotechnology.

This overview suggests several reflections on the design of industrial policy action. For instance, we can identify certain weak parts of the network that would have to be strengthened if the total network is to develop satisfactorily. This could involve reinforcing resources in certain types of units – research institutes, R&D companies, or industrial companies – or helping to build up new units. It could also be a question of stimulating and facilitating the formation of certain desirable collaborative relationships between existing actors, in Sweden or abroad. One concrete question concerning the development of Swedish biotechnology, for example, concerns the role that KabiGen (or some other genetic engineering R&D company) should play in a national research strategy. If KabiGen is to be able to make a substantial contribution to industrial development, it will probably first be necessary to furnish the company with new and greater resources. The next question concerns how this should be done. And if it is to be done, then KabiGen would also have to establish broader collaboration with the Swedish biotechnology industry, rather than concentrating almost exclusively on KabiVitrum as it does at present.

Another concrete but much more difficult question concerns relations with the international biotechnological network. By way of analogy with the individual company, national states also need to draw as much advantage as possible from developments in other countries. But we have also seen that if the individual company is to benefit to the full from the available networks, it must have internal resources of a kind to attract potential partners. The same is certainly true of the biotechnology field in Sweden.

Implanting research programmes in industry

An important part of STU's operations consists of supporting knowledge – developing and building up the technological research capability of the universities and similar institutions. Its goal is to create a knowledge base for the future development of industrial processes, methods, and products, in areas judged to have strategic importance to industrial development in Sweden.

A large part of STU's aid has been provided since the beginning of the 1980s in the form of general programmes. A general programme consists of co-ordinated research projects in some particular knowledge area. The programmes are of fairly long duration (four to seven years) and are normally carried out at several universities and research institutes.

Since the results of this research in the shape of new knowledge are to furnish a base for technological development in the related companies, the dissemination of knowledge is an important problem. If industry is to be able to benefit from this new knowledge, it must be aware of what research is going on, and there must be efficient channels for the actual dissemination of the results. The aim of the study referred to above was to investigate these two aspects by describing and analysing them with reference to six general programmes.

Three to five personal interviews and four to eleven telephone interviews were arranged for each programme with officers of STU, researchers, and representatives of industry.

Table 11.1 provides a brief summary of the way different channels were used in the six general programmes. The terminology is derived from Figure 10.4, which illustrates the theoretical framework of the study. There proved to be large differences, which was hardly surprising. The general programmes differed in several important respects which affected conditions and needs in disseminating information and knowledge. I shall use three main classes of variable to analyse and explain the observed variations. These are the content of the general programme; the nature of the knowledge networks; and the chosen method of information (and knowledge) transfer.

Content of the general programmes

One important variable in the content of the general programmes concerned the focus of the research. Although all the programmes were geared to the development of knowledge as opposed to technological development, it was none the less possible to distinguish some that were more concerned with basic knowledge from those with application. The first kind of programme starts from

Table 11.1 Frequency and importance of different channels to the dissemination of information and knowledge in the six general programmes

Programme	One-way channels		Interactive channels
	Indirect	Direct	
Electronic and electro-optic component technology	Little importance except for recruitment	Occurs rather rarely	All forms occur to a considerable extent
Information-processing	Occurs very rarely	Up to now great store has been set in this type of channel	Hitherto employed only to a limited extent by a few laboratories. However, there are discussions about starting new collaborative projects
Industrial food processors in in a nutrition perspective	Publications have been important to the dissemination of information	There have been several conferences, symposia, and other meetings	Has been of very little importance except in a couple of cases
Laser processing	Publications have been used, and will be used, as an important channel for creating interest	There have been several symposia and courses	Commissions for individual companies have been used as an important channel for the transfer of knowledge
Technology of cellulose- and fibre-based materials	There has been some information published in scientific publications, but this has meant relatively little	Hitherto only one small conference. A large symposium is planned for 1993	Collaborative research, established relationships, and the use of researchers as consultants have all occurred
Wood materials	Very little importance (one paper)	Five information meetings of various kinds have been held. The reference group includes five representatives from the wood industry	There are established relationships with the chemical companies, partly because these have formally employed researchers. Contacts with the wood companies are still at a preliminary stage

Source: Håkansson, Laage-Hellman, Axelsson (1983)

a particular area of knowledge and aims at some specific field of application or some clearly defined target group. The second kind is orientated more firmly towards problems, often in a particular industry or type of company. Of course, a single programme may contain projects of varying orientation.

The focus of the research is often connected with the stage of development which the particular research area has reached. The older the area, and the more firmly established it is in industry, the greater the tendency for research to concentrate on application. Any area of knowledge tends to split successively into a growing number of more or less clearly defined and independent research areas aiming at different or perhaps common target groups. Let us look at the area of semiconductor physics. Alongside the primary research in this field today we can distinguish several application-orientated research areas, such as process technology for integrated circuits, fibre-optic communications, and computer-aided design (CAD).

The problems connected with the dissemination of information vary with the focus of the research. When the emphasis is on the development of basic knowledge there is rarely any obvious user group (compare, for example, the general programme for information-processing). The gap between research results and industrial application is also quite wide, which means that it is generally more difficult to transfer knowledge to industry than it is in the more application-orientated programmes (such as electronics).

Another aspect of the general programme which may be significant in this context is its structure, that is how the programme is divided into different projects and where these are located. In theory we can envisage designing a general programme in one of three ways:

1. Horizontal specialization. Different research units concentrate on different parts of the field of knowledge. Research can be conducted in projects that are relatively independent of one another, each making its contribution to the overall development of knowledge in that area. Co-ordination and knowledge exchange are conducted mainly to avoid any duplication of research work.
2. Vertical specialization. Different projects that are dependent on one another are located in different research units. Co-operation between the institutes is needed to co-ordinate their activities and to ease the transfer of results from one institute to another (for instance, in the food programme analytical methods were developed at one institute and then used by other participating institutes).
3. A general programme can be designed to let several research units work on the same project at the same time. The microelectronic

group provides an example of this kind of collaboration. The advantage of such a structure is that it tends to compel better co-ordination and more efficient cross-fertilization between the various institutes. Further, the research problems can be defined in user terms, rather than on the basis of a particular knowledge area (or institute).

In real life the most common solution is probably to combine several approaches, particularly horizontal and vertical specialization. The third approach is not so usual, simply because the institutes concerned have traditionally built up their operations independently of one another. The general programmes have then largely come about by bringing together existing research groups, each of which has its own capability and special interest. Specialization often complicates the question of disseminating information, since the user side has quite a different structure, that is the problems requiring solution are not structured according to 'knowledge areas'. This calls for a good deal of hard work and effort on the part of the general programme supervisors and others.

Nature of the knowledge network

Three important features of the knowledge network are the size of the network (i.e. the number of units it includes); the capability of the member companies in the relevant knowledge area; and the existence of relations between research units and companies.

Obviously, the more units a general programme includes, the more difficult it will normally be to disseminate information and pass on results. It is also difficult if those who are to make use of the new findings possess little knowledge in the field. The problems will thus be quite different when introducing a technology that is completely alien to the target group, and when introducing new knowledge to companies that already possess considerable competence in the particular field.

Close links between research institutes and companies is essential to the efficient transfer of knowledge. If such relations already exist, the dissemination of knowledge will benefit. If there are no such relations, then channels will have to be established before any knowledge can be transferred. This is the situation in the general programme for wood materials, where for historical reasons there is practically no contact between the research chemists and the wood industry. The general programme for the technology of cellulose fibre-based materials provides an example of the opposite, with many years of well-established relations between the central research institute and the major users.

Network analysis for industrial policy

Information transfer

The methods that were chosen more or less consciously for transferring information – and consequently knowlege – in the general programmes can be described in the dimensions discussed below.

Degree of centralization/decentralization. Two quite different strategies can be distinguished here. In one case all information dissemination is controlled centrally, for example by the steering group. In the second case, no common or co-ordinated information dissemination takes place at all, and responsibility for it is placed exclusively on the individual research institutes. In most real-world cases there is a mix, in which there may be more or less emphasis on any one direction, and in which the distribution of roles can change from one time to another. At the start of a general programme and perhaps sometimes at a later stage, too, it may be appropriate to make a joint effort to inform the industry about the general programme. But in transferring and utilizing the research results, it is difficult to envisage this taking place without some interactive channels established directly between researchers and users.

When information dissemination is decentralized, several approaches are possible and the choice will be an individual one, probably depending on the different focus of research or the different nature of the network.

Level of activity. Different amounts of resources can be invested in information activities. There might be a conscious strategy to hold back at the beginning of a programme, perhaps to allow a reasonable number of results to accumulate before approaching the industry. In my view there is a considerable risk in this. When the results finally begin to be presented, they may not fit in with industry's own development plans and will therefore not be used. In any case it may be necessary to prepare industry for what is to come, and to start preparing interactive channels (if no channels exist). Another reason for 'lying low' may of course be that the research operation is itself in a start-up phase and the researchers simply have no time for external activities. Since the general programmes are all relatively new, several research groups may have been in just this situation. During an initiation phase there may thus be good reason to place the main responsibility for general information dissemination on the steering group and STU.

Timing. This dimension is closely related to the level of activity. However, it is not only the level of activity that can vary over time;

even the mix of communication channels can change. The typical pattern is to start by concentrating mainly on one-way channels, then change over to interactive channels later; in fact the first moves along the one-way channels may turn out to be one step towards the creation of interactive channels. Steering groups or reference groups can provide a suitable entry point to industry for researchers lacking contacts of their own with suitable companies. We found several examples of this in our study (e.g. in the general programme for information-processing and laser machining).

Degree of selectivity. Information can be spread broadly to as many people as possible, or to a more select target group. In the extreme case a research institute may aim at a single company (in fibre optics, for example, the Microwave Institute has communicated almost exclusively with the Eriksson Group).

Nature of the channels. The chosen method for disseminating information in any one situation will be influenced by the dimensions we have discussed. Decentralized selective information dissemination can be served best by interactive channels. But the availability of suitable channels is also an important factor: people will naturally use such channels as are available first even though they may not be the most efficient for the purpose. For example, if indirect one-way channels are available in the shape of industrial or professional journals, they should be used, although in the long run the desired effect will call for interactive channels.

Our study of the general programmes led us to a number of conclusions, some of which have been touched upon above. Three points in particular were noted. First, the efficient transfer of knowledge calls for channels of a high quality, and investment is needed on the research side and in industry for their creation. The importance of the channels varies at different times in the individual research programmes and in relation to different partners.

Second, communication channels varied between the different networks. In some cases well-developed channels already existed, while in others they were embryonic. The latter case is not the only one that can be problematic; it is sometimes necessary to communicate knowledge in a way that does not agree with the structure of the network, in which case the well-established channels themselves may be a problem. Regardless of the nature of the existing network, the dissemination of information and knowledge will thus have to be adjusted in some respects to fit it.

Our third point concerned the general lack of awareness of the need for different channels and of the need to adapt to the nature of the existing network. There were a few exceptions to the general lack of understanding on this point, but by and large conduct on the research side and in industry was fairly amateurish in this respect.

Technology exchange in the mechanical engineering industry in Uppsala County

During 1986 the consultancy group to which I belong was asked by the Uppsala County Council to conduct a study of technology exchange in the mechanical engineering industry in the county. We focused particularly on the educational aspect and on the need for technological centres. The purpose of the study was to survey the present state of technology exchange and to consider the potential contribution of a local technological centre and the needs it might be able to answer. We were also asked to consider how such a centre should be designed. What should it include? Generally speaking, the politicians seemed to imagine anything from a minor educational unit to a powerful development centre complete with specialists and advanced equipment. To get a better basis for our recommendations we were to talk to companies and try to find out how they thought they might use various types of technology centre.

Twenty companies employing between 36 and 1,700 people were included in the investigation. Altogether the companies employed 5,235 people, or about 75 per cent of the total number employed in the chosen industry. This corresponds to 34 per cent of the total number employed in industry in the region as a whole. Thus the companies represented the predominant operations within the defined technological area, as well as a considerable proportion of the operations in the region.

The study was conducted by interview. Between one and four people were interviewed in each company; in the smaller ones the manager and in the larger ones top development and production people were interviewed as well. During the interviews we identified the companies' most important technological partners inside and outside the region, and discussed questions of product and production design. We also asked about the educational background of the various personnel categories, and what the companies' ideas were about educational requirements for the future.

The network

The picture of the existing network that emerged from the interviews is illustrated in Figure 11.3. It appears that the 20 companies had some mutual ties as well as ties with other units within the region, but

Network analysis for industrial policy

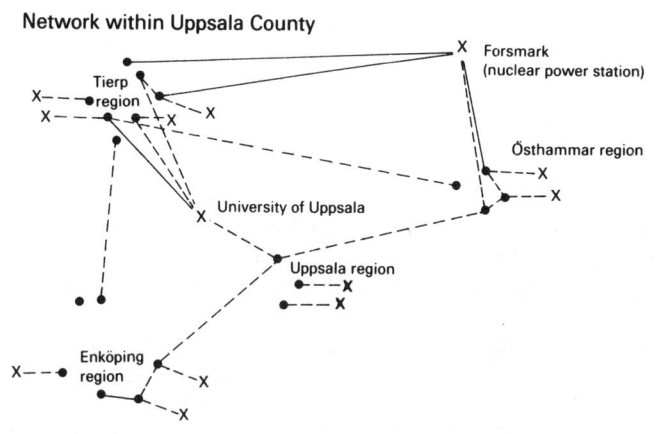

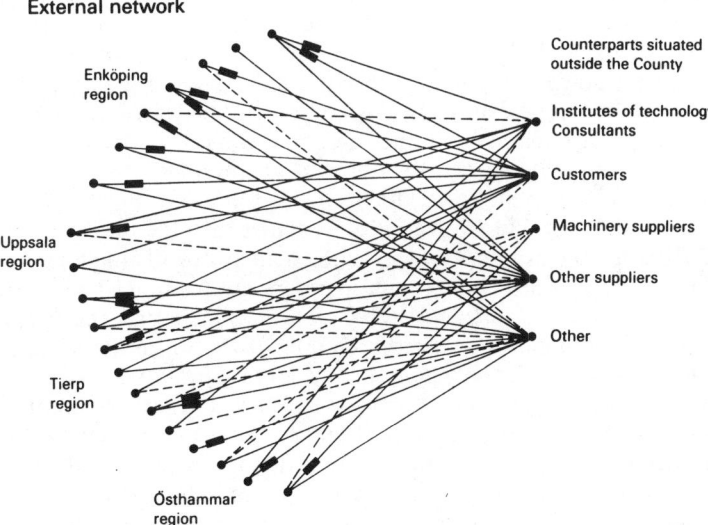

Key:
- • companies studied (20)
- X other companies in the county
- --- relationships with marginal technological content
- ―― relationships with significant technological content
- ―■― relationships with very significant technological content

Figure 11.3 The mechanical engineering network in Uppsala County in a technological development perspective

Table 11.2 Relationships with a technological content and location of the partner (number of relationships)

Location of partner	Technical content of the relationship		
	Marginal	*Significant*	*Very significant*
Inside the region	22	5	0
Outside the region	Several hundreds	About 25	18

in no case were these relationships described as 'very significant'. All 18 relationships that were classified in these terms were with partners outside the region. Table 11.2 shows the distribution of different types of relationship and the location of the partners.

We found that the 20 companies each belong to different networks, and that they therefore give one another little support, thus the regional network is not well developed.

Starting from a general network picture we can also describe the role and identity of individual units in the network. The two units University of Uppsala and Forsmark nuclear power station in Figure 11.3, for example, play only a minor role. This is particularly interesting since we know that both units possess considerable resources in the field in question, and particular efforts have been made to get the companies to do more to exploit these resources.

In sum, this network reveals some small embryonic clusters within the geographical area, based either on geographical proximity or on some special unit, but the technical exchange between these clusters has hitherto been slight. Contact with partners outside the region is very dispersed: only two external partners are mentioned by more than one company, namely Volvo as a customer and the Royal College of Technology as a research unit. It is also in relationships with partners outside the region that the more substantial technology exchanges and technological development occur. Furthermore, it is more common in these cases that the companies in the region acquire knowledge from their external partners than the other way round.

One reason for this is that about 40 per cent of the companies included in the study are satellite companies, that is production units belonging to some bigger group. A good deal of the R&D of the larger companies is conducted outside this geographical region. The situation in the regional network is summarized below:

- Technology exchange within the region is marginal.
- Relationships with partners outside the region are very fragmented and have few common denominators.

- The companies included in the study play a peripheral part in the Swedish mechanical engineering technology network as a whole.

Technological centres

The term 'technological centre' can mean almost all things to all men. The many technological centres in Sweden range from modest government-sponsored vocational training centres to major undertakings employing 10 to 20 highly qualified engineers or other specialists.

In discussing this question with the companies we quickly found that interest focused on either of the two extreme forms, that is on small local units or on large-scale technology-intensive organizations commanding a much greater geographical area. The first, according to the companies, would constitute a local resource and could be an interesting partner, particularly in connection with technological training for employees. The smaller companies could also envisage this type of unit providing computer and economic competence as well, and thus becoming a useful business centre. No-one saw the units as having any independent role in technological development; however, they could well function as a sort of information centre on certain technological questions. The smaller companies were more inclined to consider a wider range of uses for such units while the larger companies felt that in their own case the units would have their chief role in training.

The larger technology-intensive version would fulfil quite a different function in the network. It could act as a flagship, creating a special regional profile, at least within the particular technological field. It takes a long time to establish and launch organizations of this kind, and it requires great persistence on the part of those who are to manage and develop it as well as its financial backers. The core of such centres lies in development rather than training. If this type of unit is to be interesting in the long term, it must maintain relations with discriminating customers, competent suppliers, and highly qualified research units. In other words, a centre of this kind would more or less automatically turn its attention outside the region. If it is to act as a flagship, such contacts are obviously necessary. But they should not be allowed to dominate, or the link-up with the local technology network will be too limited, and the technological centre will lack the integrative function that the county council desired.

This brief description of both types of technological centre – the small and the larger – gives some indication of the dilemma facing social planners who want to influence a network. It is fairly easy to find ways of giving marginal support to existing networks, or to encourage tentative efforts towards forming networks. But if the aim is to generate any more substantial changes, then major long-term investments are needed, and they will be subject to great uncertainty. If the investment is concentrated in one area, it will be possible to create a unit strong enough to exert an active influence on the network; at the same time, however, the unit may quickly adjust to the network and become incorporated in its structure, thus tending to reinforce existing conditions rather than breaking them down. On the other hand, if the investment is divided among several smaller units, there will have to be some sort of co-ordination before the units can have any combined effect on the network.

In Uppsala County the decision was to invest primarily in fairly small technological centres, with a strong local connection and an emphasis on training.

Advantages of the network approach

The projects described above provide examples of how industrial policy issues can be defined and characterized in a network perspective. Several characteristic aspects of the network approach emerged, and in concluding this book I shall discuss the three points that seem to me to be the most important. First, a network analysis focuses on events and developments between companies and organizations rather than – as more traditional analysis is wont to do – looking chiefly at events within the companies. Consequently, the function and value of a particular company in a societal perspective is defined not only in terms of its own nature but also in light of its relationships with other companies or organizations.

Second, in a network perspective the emphasis is on specific relationships, as opposed to the more usual focus on general relations between international, national, and regional development, for instance, whereby developments in the larger or higher-level system provide the framework for developments in the smaller or lower-level system. Naturally, this type of general relationship is included in the network analysis, but an attempt is also made to look for specific relationships that link particular international conditions, for example, to the national and/or the regional. After all, it may well be some specific international relationships that determine certain events in the region and therefore make an impact on the local structure.

The third characteristic of the network analysis is the dynamic element. A network is never stable or in balance, but is always changing in all kinds of ways.

Let us now examine each of these characteristic features of the network approach and illustrate them with examples from the projects described above.

By definition, a network analysis means that what happens between companies is regarded as just as important as what happens within them. Thus in a network approach it is important to examine and influence exchanges and collaborative arrangements of various kinds. Further, there is more point in defining and characterizing actors in relation to one another than in describing them in absolute terms. The developmental thrust is seen primarily as connected with interactions between actors rather than as the sum of their internal efforts. Finally, the network approach always presents a picture of abundance and variety in the endless number of possible combinations of resources, activities, and actors; at the same time it provides a way, at least to some extent, of understanding and handling this complexity.

All three projects illustrate this point very clearly. In all three interest was focused on what was happening between companies and on the way this affected long-term development. In particular, we could see how important it is for companies to maintain active relations with the research world, if they are to be able to exploit current research effectively. Well-developed relationships help organisations to utilize the strengths and potential of other organizations, and this in turn enhances the effects of their own efforts.

A natural consequence of a network approach is to direct interest mainly towards specific relationships and dependencies between actors or parts of the network, or indeed between different networks. In this it differs from an analysis dictated by market theory, which is primarily concerned with general dependency relationships. The specific relations recognize no boundaries, either industrial, regional, or national. Nor are they all the same type, or have similar contents; on the contrary they are varied and heterogeneous. In an analytical perspective it is always dangerous to rely on internal or restricted explanatory factors. Rather, the network or the part of a network that is being analysed should be seen as an element in a larger pattern; this element has specific links with other parts of the same pattern. In the network approach it is assumed that these specific links will affect the whole or part of the network's functioning, and it is therefore important to understand them. Thus any individual relationship may be important to the functioning of the network, but it will never be decisive on its own (if it were, there would by definition be no network). Changes in any one of these important couplings may have major consequences, but there are also generally opportunities for adapting to the new situation by making changes in the other links. For these reasons network analysers serve more than the purposes of analysis; they can also be used creatively, for example in helping us to understand and recognize how different technologies or resources are – or can be – related to one another.

The biotechnology study provides an excellent example of the importance of specific relationships. The most important question for industrial policy is not whether there is a certain kind of biotechnological competence in Sweden, but how such competence as does exist is used and made to work. Just as with any other natural resource, a supply of competence is of no importance if it has no qualified users.

The third characteristic feature of the network approach is its dynamic element. In this approach the significance of various patterns of change becomes more evident. In Chapter 3 we described these patterns in terms of network processes such as heterogenization and hierarchization.

Networks are always changing because of their relations with other networks and because of internal pressures, mainly in the form of links between the different actors. It is thus assumed that, unlike the perfect competition of the market structure, for example, the network structure implies in itself a force for change.

In our projects this dynamic element was most apparent in the analysis of the genetic engineering network and the regional technology network in Uppsala County. The structure we observed in the course of the analysis was regarded in both cases as the result of earlier development and as an indicator of the continuing forces of change which would affect future developments.

These three characteristic aspects of the network approach – the consideration of events and relations between companies, the importance of specific dependence relationships to the development of larger systems, and the dynamic element – are certainly crucial to many industrial policy issues. There seems no doubt at all that the approach will prove not only useful but vital as an analytical tool in the context of industrial policy.

Appendix:

Questionnaire used in company interviews

Name of the company:

Location:

County:

    Respondent   Occupation
1.
2.
3.
4.
5.
6.

I. General characteristics

1. Company number
2. Industry code
3. Number employed
   - 1 = 20–49
   - 2 = 50–99
   - 3 = 100–149
   - 4 = 150–199
   - 5 = 200–299
   - 6 = 300–399
   - 7 = 400–499
   - 8 = ≥ 500
4. Number of man-years (same alternatives as in Q. 3)

5. Proportion white-collar workers
   - 1 = 1–9%
   - 2 = 10–19%
   - 3 = 20–29%
   - 4 = 30–39%
   - 5 = 40–49%

Appendix

    6 = 50–59%
    7 = 60–69%
    8 = ≥ 70%

6. Turnover per employed
   1 = 0–99,000 SEK   5 = 300–399,000
   2 = 100–199,000   6 = 400–499,000
   3 = 200–249,000   7 = 500–599,000
   4 = 250–299,000   8 = ≥ 600,000

7. Turnover (million SEK)
   1 = 10–24
   2 = 25–49
   3 = 50–74
   4 = 75–99
   5 = 100–149
   6 = 150–199
   7 = ≥ 200-

8. Age of company
   1 = 1–2 years
   2 = 3–4 years
   3 = 5–9 years
   4 = 10–14 years
   5 = 15–24 years
   6 = ≥ 25 years

9. Last major production change (e.g. new plant same alternatives as in Q. 8)

10. Location of the company
    1 = city (over 50,000 inhabitants)
    2 = city (less than 50,000 inhabitants)
    3 = town
    4 = countryside

11. Location of the company – kilometres to Stockholm

12. Ownership
    1 = wholly independent Swedish-owned company
    2 = independent division or subsidiary
    3 = independent foreign-owned company (more than 50% owned by foreign company or person)
    4 = production unit within Swedish group
    5 = production unit within foreign group
    6 = other type (specify)

Appendix

II. Purchasing

Share of volume for the following product groups as percentage of total purchasing volume.

	1–9%	10–24%	25–49%	50–74%	75–100%
Raw and process materials					
Components					
Supplies					
Equipment					

13. Share of raw and processed materials in the total purchased volume
    1 = 0–9%
    2 = 10–24%
    3 = 25–49%
    4 = 50–74%
    5 = 75–100%

14. Share of components (same alternatives as in Q. 13)

15. Share of supplies (same alternatives as in Q. 13)

16. Share of equipment (same alternatives as in Q. 13)

17. Total purchased volume per year in relation to total turnover
    1 = 0–24%   5 = 60–69%
    2 = 25–39%   6 = 70–79%
    3 = 40–49%   7 = 80–89%
    4 = 50–59%   8 = 90–100%

18. Largest single input product (specify product) has the following share of the total purchased volume (excluding energy or transports)
    1 = 1–9%   5 = 40–49%
    2 = 10–19%   6 = 50–59%
    3 = 20–29%   7 = 60–69%
    4 = 30–39%   8 = ⩾ 70%

Appendix

19. The second largest input product (specify product) has the following share (same alternatives as in Q. 18)

20. The third largest input product (specify product) has the following share (same alternatives as in Q. 18)

21. Regional suppliers' (within the same county or within 50 km) share of total purchased volume (same alternatives as in Q. 18)

22. The ten largest suppliers' share of total purchased volume
    1 = 1–14%   5 = 50–59%
    2 = 15–29%   6 = 60–69%
    3 = 30–39%   7 = 70–79%
    4 = 40–49%   8 = 80–100%

23. The five largest suppliers' share of the total purchased volume (same alternatives as in Q. 22)

24. The largest suppliers' share
    1 = 1–4%   5 = 20–29%
    2 = 5–9%   6 = 30–39%
    3 = 10–14%   7 = 40–49%
    4 = 15–19%   8 = $\geq$ 50%
    (Specify: _____)

25. How many of the ten largest suppliers were used five years ago?
    1 = 0   4 = 5–6
    2 = 1–2   5 = 7–8
    3 = 3–4   6 = 9–10

26. How many of the competitors of the company buy from exactly the same suppliers?
    1 = none
    2 = some
    3 = most of them
    4 = all

27. How many suppliers are there to choose among for the single largest input product?
    1 = 1   4 = 5–7
    2 = 2   5 = 8–12
    3 = 3–4   6 = > 12

Appendix

28. How many suppliers are there to choose among for the second largest input product (same alternatives as in Q. 27)?

29. How many suppliers are there to choose among for the third largest input product (same alternatives as in Q. 27)?

30. How many suppliers are there to choose among for the most important input product (specify product) (same alternatives as in Q. 27)?

31. How many of your suppliers are important from a technical development view?
    1 = 0   4 = 6–10
    2 = 1–2   5 = > 10
    3 = 3–5

    Where are these situated?

	number
regional	
Sweden	
Nordic countries	
Europe	

32. Number of suppliers important from a technical development view that are situated within the region (for 1–7 give the number; if 8 or more answer 8)

33. Same question as (32) but for suppliers situated outside the region but within Sweden

34. Same question as (32) but situated outside Sweden but in the Nordic countries

35. Same question as (32) but situated in Europe

36. Same question as (32) but situated outside Europe

How dynamic are the different supply markets in a five-year perspective regarding
a) number of suppliers (mark with an x)
b) technical development (mark with an o)
c) cyclical variation (mark with a y)

Appendix

	Stable	Small changes	Some changes	Relatively large changes	Very large changes
	1	2	3	4	5
Raw and processed materials					
Components					
Equipment					

37. Pattern of change regarding number of suppliers of raw and processed materials (alternatives as in 1–5 above)

38. Pattern of change regarding technical development of the input markets for raw and processed materials (alternatives 1–5)

39. Pattern of change regarding cyclical changes of input markets of raw and processed materials (alternatives 1–5)

40. Pattern of change regarding the number of suppliers of components (alternatives 1–5)

41. Pattern of change regarding technical development of input markets for components (alternatives 1–5)

42. Pattern of change regarding cyclical variations of input markets for components (alternatives 1–5)

43. Pattern of change regarding the number of suppliers of equipment (alternatives 1–5)

44. Pattern of change regarding technical development of input markets for equipment (alternatives 1–5)

45. Pattern of change regarding cyclical variations of input markets for equipment (alternatives 1–5)

Appendix

How important is your company to the 10 largest suppliers in your opinion?

	Volume No. of the 10	Technical development No. of the 10
Very important		
Rather important		
Rather marginal		
Completely marginal		

46. The investigated company is very important in terms of volume for_____ suppliers (if 8 or more write 8)

47. The investigated company is rather important in terms of volume for_____ suppliers

48. The investigated company is rather marginal in terms of volume for_____ suppliers

49. The investigated company is completely marginal in terms of volume for_____ suppliers

50–53. Same questions as 46–49 but in regard to technical development

54. Import share of the purchases
    1 = 0%   5 = 35–49%
    2 = 1–9%   6 = 50–74%
    3 = 10–24%   7 = 75–100%
    4 = 25–34%

55. Share of the total imports coming from the Nordic countries
    1 = 0%   5 = 50–64%
    2 = 1–19%   6 = 65–79%
    3 = 20–34%   7 = 80–100%
    4 = 35–49%

56. Share from Europe apart from the Nordic countries (same alternatives as in Q. 55)

57. Share from countries outside Europe (same alternatives as in Q. 55)

Appendix

58. The share of imports has during the last five years
    1 = decreased sharply (sharply is defined as at least 10%)
    2 = decreased somewhat
    3 = been unchanged
    4 = increased somewhat
    5 = increased sharply

59a. Does the company control any input resources through ownership? Estimate the share of total purchases
    1 = 0%   5 = 20–29%
    2 = 1–4%   6 = 30–39%
    3 = 5–9%   7 = ≥ 40%
    4 = 10–19%

New products' share of total volume of purchase in a five-year perspective (only more radical changes)

59b. New raw or processed materials' share of total volume (same alternatives as in Q. 59a)

60. New components' share of total volume (same alternatives as in Q. 59)

61. New equipments' share of total volume (same alternatives as in Q. 59)

Three most important supplier relationships from a technical development point of view: Supplier 1 _____

62. Type of product
    1 = raw or processed material
    2 = component
    3 = equipment
    4 = other (specify)

63. Location of supplier
    1 = within the same town   4 = within the Nordic countries
    2 = within the county or   5 = within Europe
        50 kilometres   6 = U.S. or Canada
    3 = within Sweden   7 = rest of the world (specify)

Appendix

64. Age of the relationship
    1 = 0–1.9 years
    2 = 2–4.9 years
    3 = 5–9.9 years
    4 = 10–14.9 years
    5 = 15–19.9 years
    6 = 20–29.9 years
    7 = over 30 years

65. Purchased volume as percentage of total volume purchased
    1 = 0–0.9%
    2 = 1–2.9%
    3 = 3–4.9%
    4 = 5–7.4%
    5 = 7.5–9.9%
    6 = 10–14.9%
    7 = 15–24.9%
    8 = $\geq 25\%$

66. Form of relationship
    1 = part of an ongoing relationship
    2 = yearly agreement
    3 = long-range agreement
    4 = joint company
    5 = other form (specify)

67. Number of persons involved on the supplier's side
    1 = 1
    2 = 2–3
    3 = 4–6
    4 = 7–9
    5 = 10–14
    6 = ≥ 15

68. Number of persons in the investigated company who are involved in the relationship (same alternatives as in Q. 67)

69. Frequency of personal contacts
    1 = once a year
    2 = once every half-year
    3 = once every third month
    4 = once every month
    5 = once every fortnight
    6 = once every week
    7 = several times every week

70. Type of development co-operation
    1 = buying on special terms
    2 = joint technological information exchange
    3 = tests etc.
    4 = completed a special development project
    5 = joint development work – team
    6 = long-range technical co-operation

Appendix

71. The relationships have up to now for the investigated company resulted in
    1 = nothing concrete
    2 = a slight improvement
    3 = several slight improvements
    4 = a major improvement
    5 = several major improvements

72. The results for the supplier (same alternatives as in Q. 71)

73. Our future expectations (same alternatives as in Q. 71)

Supplier 2 _____

74–85. Same questions as 62–73

Supplier 3 _____

86–97. Same questions as 62–73.

III. Selling side

98. Number of important product groups (each with at least 5% of the total turnover) (if more than 8 write 8)

99. The largest product group stands for _____ % of the turnover
    1 = 1–19%   4 = 60–79%
    2 = 20–39%   5 = 80–100%
    3 = 40–59%

100. The second largest product group stands for (same alternatives as in Q. 99)

101. The third largest product group stands for (same alternatives as in Q. 99)

102. The different product groups are sold to
    1 = the same buyers
    2 = mainly the same buyers
    3 = mainly different types of buyer
    4 = quite different types of buyer

Appendix

103. A first-time user of a product from your most important product group can use the product
    1 = without any problems
    2 = with some introductory instructions
    3 = with extensive introductory instructions
    4 = with introduction and day-to-day advice

104. Total number of buyers (if more than 10,000 write 9,999)

105. The biggest buyer's share of total sales
    1 = 1–4%   5 = 20–29%
    2 = 5–9%   6 = 30–39%
    3 = 10–14%   7 = 40–49%
    4 = 15–19%   8 = ⩾ 50%

106. The five biggest buyers' share
    1 = 1–14%   5 = 50–59%
    2 = 15–29%   6 = 60–69%
    3 = 30–39%   7 = 70–79%
    4 = 40–49%   8 = ⩾ 80%

107. The ten biggest buyers' share (same alternatives as in Q. 106)

108. Regional buyers' share
    1 = 0–4%   5 = 30–39%
    2 = 5–9%   6 = 40–49%
    3 = 10–19%   7 = 50–74%
    4 = 20–29%   8 = ⩾ 75%

109. Potential number of buyers (excluding existing customers) in Sweden
    1 = very few   4 = many
    2 = few   5 = uncountable
    3 = some

110. Potential number of buyers worldwide (same alternatives as in Q. 109)

111. How many of those now among the ten biggest buyers were among the 50–60 biggest customers five years ago?

    1 = 0   4 = 5–6
    2 = 1–2   6 = 7–8
    3 = 3–4   7 = 9–10

183

Appendix

112. How many of the ten biggest buyers are buying from your biggest competitor (same alternatives as in Q. 111)?

113. Are the ten biggest buyers similar in regard to their buying patterns (knowledge, development, and production characteristics)?
     1 = very similar
     2 = marginal variation
     3 = some variation
     4 = rather different
     5 = very different

114. The ten biggest buyers are treated from a technological point of view
     1 = very similarly
     2 = rather similarly
     3 = with some differentiation
     4 = rather differently
     5 = very differently

How important are you to the ten biggest buying companies (including distributors)

	Very important	Rather important	Some importance	Rather marginal	Completely marginal
Importance in terms of volume					
Importance in terms of technical development					

Questions 115 to 119 concern your importance in terms of volume.

115. Very important for _____ buyers (if 8 or more write 8)

116. Rather important for _____ buyers

117. Some importance for _____ buyers

118. Rather marginal for _____ buyers

119. Completely marginal for _____ buyers

Appendix

Questions 120 to 124 concern your importance from a technical development point of view.

120. Very important for _____ buyers

121. Rather important for _____ buyers

122. Some importance for _____ buyers

123. Rather marginal for _____ buyers

124. Completely marginal for _____ buyers

125. How big a share of the sales in Sweden goes to distributors?
 1 = 0%   5 = 40–49%
 2 = 1–14%   6 = 50–59%
 3 = 15–29% 7 = 60–74%
 4 = 30–39% 8 = 75–100%

126. How big a share of the export goes to distributors (same alternatives as in Q. 125)

127. Export share of total sales
 1 = 0%   4 = 25–49%
 2 = 1–9%   5 = 50–74%
 3 = 10–24%   6 = ≥ 75%

 Export distribution:

Market	Export share (%)
Nordic countries	_____
Europe	_____
_____	_____
_____	100%

128. Share of exports going to the Nordic countries (same alternatives as in Q. 127)

129. Share of exports going to Europe (outside Nordic countries) (same alternatives as in Q. 127)

130. Share of exports going to the rest of the world (same alternatives as in Q. 127)

Appendix

The exports dominated by the following numbers and types of customer:								
	Number				Type			
	1	2–4	5–10	>10	Distrib.	OEM	End-users	combinations
Nordic								
Europe								
Other countries								

131. The exports to the Nordic countries are dominated by the following numbers of buyers
    1 = 1   3 = 5–10
    2 = 2–4   4 = > 10

132. The exports to Europe (outside Nordic countries) are dominated by (same as in Q. 131)

133. The exports to North America are dominated by (same as in Q. 131)

134. The exports to the remaining countries are dominated by (same as in Q. 131)

135. The buyers in the Nordic countries are mainly
    1 = distributors   3 = end-users
    2 = OEM-manufacturers   4 = a combination of 1–3

136. Buyers for the rest of the exports are mainly (same as in Q. 135)

137. In terms of volume exports have during the last five years
    1 = decreased heavily
    2 = decreased somewhat
    3 = been steady
    4 = increased somewhat
    5 = increased heavily
        (heavily = 10% or more)

138. Total sales costs (including technical adaptation costs) in relation to sales price are
    1 = 0%   5 = 30–39%
    2 = 5–9%   6 = 40–59%
    3 = 10–19%   7 = 60–74%
    4 = 20–29%   8 = ≥ 75%

Appendix

139. Number of competitors in the Swedish market
 1 = 0   4 = 5–9
 2 = 1–2   5 = 10–14
 3 = 3–4   6 = ≥ 15

140. Number of competitors in the export markets (same as in Q. 139 plus 7 = big variation)

141. The largest competitor's share in Sweden (same as in Q. 138)

142. The largest competitor's share in total (the world) (same as in Q. 138)

143. The largest competitor is in what relation to your company as regards market share
 1 = much bigger   4 = somewhat smaller
 2 = somewhat bigger   5 = much smaller
 3 = same size

144. During the last five years the following number of competitors have arisen
 1 = 0   4 = 4–6
 2 = 1   5 = ≥ 7
 3 = 2–3

145. Number of competitors who have disappeared during the last five years (same as in (144))

146. The competitors are
 1 = very similar
 2 = rather similar
 3 = rather different
 4 = very different

147. The sales abroad
 1 = are handled by company headquarters
 2 = are handled by agents
 3 = are handled by wholly owned subsidiaries
 4 = are handled to at least 50% by own subsidiaries
 5 = other form (specify)

148. Importance of technical sales service
 1 = does not exist
 2 = the company's technicians now and then function as such

Appendix

3 = special personnel take care of these matters
4 = special personnel that are believed to be very important take care of these matters

149. The company's market share in Sweden is
    1 = 0–9%   5 = 40–49%
    2 = 10–19%   6 = 50–59%
    3 = 20–29%   7 = 60–74%
    4 = 30–39%   8 = ≥ 75%

150. The company's market share in the Nordic countries is (same as in Q. 149)

151. The company's market share in Europe is (same as in Q. 149)

152. The company's market share in the world is (same as in Q. 149)

	Market leader	Among the biggest	In the middle	Among the smallest	Smallest
Local region					
Sweden					
Nordic countries					
Europe					
Other					

153. The market situation in the local region
    1 = smallest   4 = among the bigger
    2 = among the smaller   5 = market leader
    3 = in the middle

154. The market situation in Sweden (same as in Q. 153)

155. The market situation in Nordic countries (same as in Q. 153)

156. The market situation in Europe (same as in Q. 153)

Appendix

157. The market share has during the last five years (in the main market)
    1 = decreased substantially (substantially ≥ 5%)
    2 = decreased somewhat
    3 = not changed
    4 = grown a little (little 1–4%)
    5 = grown substantially

158. Share of the turnover for new products (in a five-year perspective)
    1 = 0%   5 = 30–39%
    2 = 1–9%   6 = 40–49%
    3 = 10–19%   7 = 50–74%
    4 = 20–29%   8 = ≥ 75%

159. Share of the turnover for new products that your company developed in a five-year perspective
    1 = 0–4%   5 = 20–29%
    2 = 5–9%   6 = 30–39%
    3 = 10–14%   7 = 40–49%
    4 = 15–19%   8 = ≥ 50%

Technological development relationships exist in the following form and with following number of customers		
	with ___ number of 10 biggest	with ___ number of other customers
Continuous exchange of ideas and information		
Tests		
Development co-operation (including project group)		
Long-range co-operation		

160. Technological development co-operation in any form exists in the following number of the ten biggest customer relationships
    1 = 1   4 = 5–6
    2 = 1–2   5 = 7–8
    3 = 3–4   6 = 9–10

Appendix

161. Technological co-operation on at least the test level exists in the following number of the ten biggest relationships (same alternatives as in Q. 160)

162. Of the above (Q. 161) the following number are situated in the same region as the company
    1 = 0   5 = 4
    2 = 1   6 = 5–6
    3 = 2   7 = 7–8
    4 = 3   8 = 9–10

163. Situated in Sweden (same as in Q. 162)

164. Situated in the Nordic countries (same as in Q. 162)

165. Situated in Europe (same as in Q. 162)

166. With customers other than the ten biggest, technical co-operation on at least the test level occurs
    1 = very often
    2 = often
    3 = sometimes
    4 = rarely
    5 = never

The three most important customer relationships seen from a technical development point of view

Questions 167 to 177 concern relations with the most important buyer.

167. The buyer is situated
    1 = within the town   5 = within Europe
    2 = within the county   6 = within North America
    3 = within Sweden   7 = the rest of the world
    4 = within Nordic countries

168. Duration of the relation
    1 = 0–1.9 years   5 = 15–19.9 years
    2 = 2–4.9 years   6 = 20–29.9 years
    3 = 5–9.9 years   7 = $\geq$ 30 years
    4 = 10–14.9 years

Appendix

169. Sales volume in relation to total sales of the product group
     1 = 0–4%   5 = 20–34%
     2 = 5–9%   6 = 35–49%
     3 = 10–14%   7 = 50–74%
     4 = 15–19%   8 = 75–100%

170. Form of co-operation
     1 = part of an ongoing relation
     2 = yearly contract
     3 = long-range contract
     4 = joint company, etc.
     5 = other (specify)

171. Number of persons in the buying company who are directly involved in the relationship
     1 = 1   4 = 7–9
     2 = 2–3   5 = 10–14
     3 = 4–6   6 = ≥ 15

172. Number of persons in the selling company who are involved in the relationship (same alternatives as in Q. 171)

173. Frequency of direct contacts
     1 = 1 per year   5 = 1 per fortnight
     2 = 1 per half-year   6 = 1 per week
     3 = 1 per quarter-year   7 = several times per week
     4 = 1 per month   8 = every day

174. Type of development co-operation
     1 = close to normal sales
     2 = joint technological information exchange
     3 = tests, etc.
     4 = special technical project
     5 = joint technical co-operation
     6 = long-range technical co-operation (several projects)

175. Results of the co-operation for the focal company
     1 = nothing concrete
     2 = one small improvement
     3 = several small improvements
     4 = one large improvement
     5 = several large improvements

Appendix

176. Results of the co-operation for the buyer (same as in Q. 175)

177. Future expectations of the relationships of the focal company (same alternatives as in Q. 175)

178–188. Same questions as 167–177 but in regard to the second most important buyer

189–199. Same questions as 167–177 but in regard to the third most important buyer

IV. **Personnel**

200. Share of employers with an education of at least college level
1 = 0–4%	5 = 20–29%
2 = 5–9%	6 = 30–49%
3 = 10–14%	7 = 50–74%
4 = 15–19%	8 = 75–100%

201. Share of employers with an education of university level
1 = 0–1%	5 = 15–19%
2 = 2–4%	6 = 20–29%
3 = 5–9%	7 = 30–49%
4 = 10–14%	8 = $\geq$ 50%

202. The density of engineers – number of engineers in relation to total number of employed (same alternatives as in Q. 200)

203. Number of engineers with a university degree in relation to total number of employed (same alternatives as in Q. 200)

204. As compared with competitors, the proportion of the engineers is
    1 = much lower
    2 = somewhat lower
    3 = the same
    4 = somewhat higher
    5 = much higher

Appendix

205. Share of the personnel that during the last year took part in internal education (of at least one day duration)
    1 = 0–9%   5 = 40–49%
    2 = 10–19%   6 = 50–74%
    3 = 20–29%   7 = 75–100%
    4 = 30–39%

206. The blue-collar workers have on an average worked
    1 = 0–2 years   4 = 10–14 years
    2 = 3–4 years   5 = 15–24 years
    3 = 5–9 years   6 = ≥ 25 years

207. The white-collar workers have on an average worked (same alternative as in Q. 206)

208. Has the mobility of the blue-collar workers over the years
    1 = increased substantially
    2 = increased somewhat
    3 = been unchanged
    4 = decreased somewhat
    5 = decreased substantially

209. Has the mobility of the white-collar workers over the years (same as in Q. 208)

210. Is there a shortage of qualified technicians?
    1 = very much so
    2 = to a certain degree
    3 = somewhat
    4 = to a very little extent
    5 = not at all

211. Is there a shortage of qualified blue-collar workers (crafts people) (same as in Q. 210)

The level of technical competence within the company has over the last 5–10 years changed in the following way for the following functions

	Decreased somewhat	Unchanged	Increased somewhat	Increased a lot	A very large increase
Production					
Purchasing					
Marketing					
Development					

Appendix

212. The technical competence level within production has during the last ten years
 1 = decreased somewhat
 2 = been stable
 3 = increased somewhat
 4 = increased a lot
 5 = a very large increase

213. Level of technical competence in purchasing (same alternatives as in Q. 212)

214. Level of technical competence in marketing (same alternatives as in Q. 212)

215. Level of technical competence in development (same alternatives as in Q. 212)

V. Capital

216. Capital intensity 1 = the book value of construction and equipment capital divided by total capital
 1 = 0–0.09   5 = 0.40–0.49
 2 = 0.10–0.19 6 = 0.50–0.59
 3 = 0.20–0.29 7 = 0.60–0.69
 4 = 0.30–0.39 8 = $\geq$ 0.70

217. Capital intensity 2 = the value of construction and equipment capital as shown by the fire insurance divided by total capital
 1 =   0–100,000   5 =   700,000–899,000
 2 = 100,000–299,000   6 =   900,000–1,099,000
 3 = 300,000–499,000   7 = 1,100,000–1,499,000
 4 = 500,000–699,000   8 = $\geq$ 1,500,000

218. Capital intensity 3 = total investments during the last five years divided by the average number of employed (same alternatives as in Q. 217)

219 Total number of share-owners
 1 =   1   5 =   50–99
 2 =   2–3   6 =   100–999
 3 =   4–9   7 = 1,000–9,999
 4 = 10–49   8 = $\geq$ 10,000

Appendix

220. The ten greatest share-owners share
    1 = 0–9%   5 = 40–49%
    2 = 10–19%   6 = 50–69%
    3 = 20–29%   7 = 70–89%
    4 = 30–39%   8 = 90–100%

221. Long-term debts' share of the total capital (same alternatives as in Q. 220)

222. Short-term debts' share of the total capital (same alternatives as in Q. 220)

223. Own capital in relation to total capital (same alternatives as in Q. 220)

224. Is there a representative of a bank on the board?
    1 = no
    2 = yes

225. Number of bank relations
    1 = 1   4 = 4
    2 = 2   5 = 5
    3 = 3

226. When did you last change banks?
    1 = 1– 2 years ago   4 = 11–15 years ago
    2 = 3– 5 years ago   5 = have only had one bank
    3 = 6–10 years ago

227. Return on investment (own capital) = profit divided by own capital
    1 =   0–0.9%   5 = 15.0–19.9%
    2 = 1.0–4.9%   6 = 20.0–24.9%
    3 = 5.0–9.9%   7 = $\geq 25\%$
    4 = 10.0–14.9%

228. Return on investment = profit divided by total capital
    1 =   0–0.9%   5 = 7.5–9.9%
    2 = 1.0–2.9%   6 = 10.0–14.9%
    3 = 3.0–4.9%   7 = 15.0–19.9%
    4 = 5.0–7.4%   8 = $\geq 20\%$

Appendix

229. Return as in Q. 227 – average for the last three years (same alternatives as in Q. 227)

230. Return as in Q. 228 – average for the last three years (same alternatives as in Q. 228)

231. Growth in volume terms during the last year
 1 = 0–4%   5 = 20–24%
 2 = 5–9%   6 = 25–29%
 3 = 10–14%   7 = 30–34%
 4 = 15–19%   8 = ≥ 35%

232. Growth in volume during the year before (same as in Q. 231)

233. Growth in volume during the year, two years before (same as in Q. 231)

234. Growth in volume during the year, three years before (same as in Q. 231)

235. Growth in volume during the year, four years ago (same as in Q. 231)

236. Volume growth on average for the last five years (same as in Q. 231)

VI. Technical features

Production technology

Give a short description:_____

_____

_____

237. Production technology
 1 = unit production
 2 = small batch production
 3 = large batch production
 4 = process production
 5 = combination

Appendix

238. Share customer-led production
    - 1 = 0–9%
    - 2 = 10–19%
    - 3 = 20–29%
    - 4 = 30–39%
    - 5 = 40–49%
    - 6 = 50–59%
    - 7 = 60–74%
    - 8 = ⩾ 75%

239. Value share = turnover – purchased products divided by turnover
    - 1 = 0–0.09
    - 2 = 0.10–0.19
    - 3 = 0.20–0.29
    - 4 = 0.30–0.39
    - 5 = 0.40–0.49
    - 6 = 0.50–0.59
    - 7 = 0.60–0.69
    - 8 = ⩾ 0.70

240. Share of total development costs that are related to product development
    - 1 = 0–9%
    - 2 = 10–19%
    - 3 = 20–29%
    - 4 = 30–39%
    - 5 = 40–49%
    - 6 = 50–59%
    - 7 = 60–74%
    - 8 = ⩾ 75%

241. Share of total development costs that are directed towards rationalizations (same alternatives as in Q. 240)

242. (deleted)

243. Sales price per kilo
    - 1 = 0–9 SCr
    - 2 = 10–49 SCr
    - 3 = 50–99 SCr
    - 4 = 100–199 SCr
    - 5 = 200–499 SCr
    - 6 = 500–999 SCr
    - 7 = 1,000–9,999 SCr
    - 8 = ⩾ 10,000 SCr

Horizontal units

244. Number of important development relationships with horizontal units
    - 1 = 1
    - 2 = 2
    - 3 = 3
    - 4 = 4
    - 5 = 5
    - 6 = 6–8
    - 7 = 9–14
    - 8 = > 15

245. Are there important horizontal units that are very active in technical development issues and with which the company has no relationship? If so, how many (same alternatives as in Q. 244)?

Appendix

246. How important is the company to the ten most important horizontal units from a development point of view?
Very important for _____ units (if more than 9, write 9)

247. Rather important for _____ units

248. Rather marginal for _____ units

249. Completely marginal for _____ units

250. Competitors have horizontal relationships with
1 = completely other units
2 = same type but other units
3 = one or two which have horizontal relationships with at least some of the same
4 = several which have horizontal relationships with the same
5 = all have contacts with at least some of those mentioned above

251. Of the five most important horizontal units the following number were involved five years ago
1 = 1   4 = 4
2 = 2   5 = 5
3 = 3

252. Is there any connection between the different horizontal units?
1 = no connection
2 = some connections but without any significance
3 = some connections of significance
4 = several connections with minor influence
5 = several connections with major influence
6 = there is a clear pattern

Three most important horizontal development relationships

Questions 253 to 262 concern relations with the most important horizontal unit.

253. This unit is located
1 = within the town   5 = within Europe
2 = within the county   6 = within North America
3 = within Sweden   7 = rest of the world (specify)
4 = within Nordic countries

Appendix

254. Age of the relation
     1 = 0–1.9 years   5 = 15–19.9 years
     2 = 2–4.9 years   6 = 20–29.9 years
     3 = 5–9.9 years   7 = $\geq$ 30 years
     4 = 10–14.9 years

255. Co-operation form
     1 = informal
     2 = yearly contract
     3 = long-range contract
     4 = joint compnay
     5 = another form (specify)

256. Number of persons in the horizontal unit who take part in the relationship
     1 = 1   4 = 7–9
     2 = 2–3   5 = 10–14
     3 = 4–6   6 = $\geq$ 15

257. Number of persons in the focal company who take part in the relationship (same alternatives as in Q. 256)

258. Frequency of contacts
     1 = 1 per year   5 = 1 per fortnight
     2 = 1 per half-year   6 = 1 per week
     3 = 1 per quarter year 7 = several times per week
     4 = 1 per month   8 = every day

259. Type of development relationships
     1 = buying of a licence
     2 = joint information exchange
     3 = tests, etc.
     4 = special project
     5 = joint development work – project group
     6 = long-range technical co-operation – several projects

260. Results for the focal company
     1 = nothing concrete
     2 = one small improvement
     3 = several small improvements
     4 = one large improvement
     5 = several large improvements

261. For the counterpart (same as in Q. 260)

Appendix

262. Future expectations for the focal company (same as in Q. 260)

263–272. Same questions as 253–262 but in regard to the second most important horizontal unit.

273–282. Same questions as 253–262 but in regard to the third most important horizontal unit.

Development work

Which persons and functions within the company take part in the development work? How many man-years?	Concrete development work	Information exchange
General management		
Development department, etc.		
Production		
Marketing		
Purchasing		
Other departments		
Total number of man-years		

283. Total number of man-years invested in technical development divided by total number of man-years
 1 =   0–0.009   5 = 0.060–0.079
 2 = 0.010–0.019   6 = 0.080–0.099
 3 = 0.020–0.039   7 = 0.100–0.199
 4 = 0.040–0.059   8 = ≥ 0.2

284. Total number of man-years invested in 'concrete' development work divided by total number of man-years invested in technical development work
 1 = 0–9%   5 = 40–49%
 2 = 10–19%   6 = 50–64%
 3 = 20–29%   7 = 65–79%
 4 = 30–39%   8 = 80–100%

285. Number of man-years invested in information exchange regarding technical development divided by the total number of man-years invested in technical development work (same alternatives as in Q. 284)

286. Share of 'concrete' developments that take part in external co-operation (same alternatives as in Q. 284)

287. Share of information exchange that involve an external partner (same alternatives as in Q. 284)

Bibliography

Alchian, A. A. and Demsetz, H. (1972) 'Production, information costs and economic organization', *American Economic Review* 5: 777-95.
Alderson, W. (1965) *Dynamic Marketing Behaviour. A Functionalist Theory of Marketing*, Homewood: Richard D. Irwin.
Allen, T. J. (1977) *Managing the Flow of Technology*, Cambridge, Mass.: MIT Press.
Axelrod, R. (1984) *The Evolution of Cooperation*, New York: Basic Books.
Axelsson, B. (1987) 'Supplier management and technological development', in H. Håkansson (ed.) *Industrial Technological Development. A Network Approach*, London: Croom Helm.
Axelsson, B. and Håkansson, H. (1979) *Wikmanshyttans uppgång och fall. En analys av ett stålföretag och dess omgivning under 75 år* (The rise and fall of Wikmanshyttan. An analysis of a steel company and its environment during 75 years), Lund: Studentlitteratur.
—— (1984) *Inköp för konkurrenskraft* (Purchasing for competitive power), Stockholm: Liber.
—— (1986) 'The development role of purchasing in an international oriented company', in P. W. Turnbull and S. Paliwoda (eds) *Research in International Marketing*, London: Croom Helm.
Benvignati, A. M. (1983) 'International technology transfer patterns in a traditional industry', *Journal of International Business Studies*, Winter 1983: 63-75.
Blau, P. M. (1964) *Exchange and Power in Social Life*, New York: John Wiley.
Blumenthal, D., Gluck, M., Louis, K. S., Stoto, M. A., and Wise, D. (1986) 'Industry research relationships in biotechnology: Implications for the university', *Science* 232 (June): 1361-6.
Cook, K. S. and Emerson, R. M. (1978) 'Power, equity and commitment in exchange networks', *American Sociological Review* 43 (October): 721-39.
Cyert, R. M. and March, J. G. (1963) *A Behavioral Theory of the Firm*, Englewood Cliffs, N.J.: Prentice Hall.
Czepiel, J. A. (1974) 'Word-of-mouth processes in the diffusion of major technological innovation', *Journal of Marketing Research* 11 (May): 172-80.

Bibliography

Dibner, M. D. (1986) 'Biotechnology in Europe', *Science* 232 (June): 1367–72.
Eccles, R. G. (1981) 'The quasi firm in the construction industry', *Journal of Economic Behavior and Organization*, 2: 335–57.
Epstein, E. M. (1969) *The Corporation in American Politics*, Englewood Cliffs, N.J.: Prentice Hall.
Evan, W. M. (1966) 'The organization-set: Toward a theory of interorganizational relations', in J. Thompson (ed.) *Approaches to Organizational Design*, Pittsburgh, P.A.: University of Pittsburgh Press.
Ford, D. I., Håkansson, H. and Johanson, J. (1986) 'How do companies interact?', *Industrial Marketing and Purchasing* 1 (1): 26–41.
Gadde, L-E. and Håkansson, H. (forthcoming) 'Analysing change and stability in distribution channels – a network approach', in B. Axelsson and G. Easton (eds) *Industrial Networks. The New Reality*, London: Routledge.
Galeano, E. (1976) *Latinamerikas öppna ådror* (Swedish translation of 'Las venas abiertas de América Latina'), Stockholm: Prisma.
Granstrand, O. (1979) *Technology, Management and Markets*, Gothenburg: Chalmers University of Technology.
Hägg, I. and Johanson, J. (eds) (1982) *Företag i nätverk – ny syn på konkurrenskraft* (Firms in networks – A new perspective of competitive power), Stockholm: SNS.
Håkansson, H. (1979) 'Marknadsföring av specialstål' (Marketing of special steel), *Research Report*, no. 2, Department of Business Administration, University of Uppsala.
Håkansson, H. (ed.) (1982) *International Marketing and Purchasing of Industrial Goods. An Interaction Approach*, Chichester: John Wiley.
Håkansson, H. (1985) 'The Swedish approach to Europe', in P. W. Turnbull and J-P. Valla (eds) *Strategies for International Industrial Marketing*, London: Croom Helm.
Håkansson, H. (ed.) (1987) *Industrial Technological Development. A Network Approach*, London: Croom Helm.
Håkansson, H. (forthcoming) 'Evolution processes in industrial networks', in B. Axelsson and G. Easton (eds) *Industrial Networks. The New Reality*, London: Routledge.
Håkansson, H. and Johanson, J. (1985) 'A model of industrial networks', Working Paper, Department of Business Administration, University of Uppsala. Revised edition to be published in B. Axelsson and G. Easton (eds) *Industrial Networks. The New Reality*, London: Routledge (forthcoming).
—— (1988) 'Formal and informal cooperation strategies in international industrial networks', in F. J. Contractor and P. Lorange (eds) *Cooperative Strategies in International Business*, Lexington, Mass.: Lexington Books.
Håkansson, H., Laage-Hellman, J., and Axelsson, B. (1983) 'Ramprogram för kunskapsutveckling – utvärdering av industriell förankring' (Research programs for knowledge development – Assessment of industrial liaisons), *STU Report 82-5372*, Stockholm: National Board for Technical Development.
Håkansson, H. and Snehota, I. (1976) *Marknadsplanering. Ett sätt att skapa*

nya problem? (Market planning. A way to create new problems?), Lund: Studentlitteratur.

Håkansson, H. and Waluszewski, A. (1986) 'Technical development in a dense network', paper presented at the Third International I.M.P. Research Seminar on International Marketing, Lyon, 3–5 September 1986.

Hamfelt, C. and Lindberg, A-K. (1987) 'Technological development and the individual's contact network', in H. Håkansson (ed.) *Industrial Technological Development. A Network Approach*, London: Croom Helm.

Hammarkvist, K-O., Håkansson, H., and Mattsson, L-G. (1982) *Marknadsföring för konkurrenskraft* (Marketing for competitive power), Malmö: Liber.

Henriksson, H. and Håkansson, H. (1985) 'Teknisk utveckling genom företag i samverkan' (Technical development through cooperations between companies), in S. Berger, K. Haraldsson, and M. Lundmark (eds) *Bostadsbyggande och byggmaterialindustrin*, Research Report no. 88, Department of Human Geography, University of Uppsala.

Hermansson, C. H. (1982) *Kapitalister* (Capitalists), Stockholm: Arbetarkultur.

Hörnell, E., Vahlne, J-E., and Wiedersheim-Paul, F. (1973) *Export och utlansetableringar* (Export and foreign establishments), Uppsala: Almqvist & Wiksell.

Jacobs, D. (1974) 'Dependency and vulnerability: An exchange approach to the control of organizations', *Administrative Science Quarterly* 19: 45–59.

Johanson, J. and Mattsson, L-G. (1985) 'Marketing investments and market investments in industrial networks', *International Journal of Research in Marketing* 2: 185–95.

—— (1988) 'Internationalisation in industrial systems – A network approach', in N. Hood and J-E. Vahlne (eds) *Strategies in Global Competition*, New York: Croom Helm.

Kagono, T., Nonaka, K., Sakakibara, K., and Okumura, A. (1985) *Strategic vs. Evolutionary Management: A U.S.–Japan Comparison of Strategy and Organization*, Amsterdam: North Holland.

Kinch, N. (forthcoming) 'Emerging strategies in a network context: The Volvo case', *Scandinavian Journal of Management Studies*.

Laage-Hellman, J. (1984) 'The role of external technical exchange in R&D. An empirical study of the Swedish special steel industry', *MTC Research Report* 18 (Stockholm).

—— (1987) 'Process innovation through technical cooperation', in H. Håkansson (ed.) *Industrial Technological Development. A Network Approach*, London: Croom Helm.

—— (in press) 'Technological development in industrial networks', Ph.D. dissertation, Department of Business, University of Uppsala.

Laage-Hellman, J. and Axelsson, B. (1986) 'Bioteknisk forskning i Sverige. Forskningsvolym, forskningsinriktning, samarbetsmönster' (Biotechnology research in Sweden. Volume, content and cooperations), *STU Report 536–1986*, Stockholm.

Bibliography

Liljegren, G. (1988) *Interdependens och dynamik i långsiktiga kundrelationer. Industriell försäljning i ett nätverksperspektiv* (Interdependency and dynamism in long-range customer relationships, dissertation with English summary), Stockholm: EFI Stockholm School of Economics.

Lindblom, C. E. (1959) 'The science of muddling through', *Public Administration Review*, American Society for Public Administration, Washington D.C., 19 (Spring): 79–88.

Lindqvist, S. (1984) *Technology on Trial. The Introduction of the Steam Power Technology into Sweden 1715–1736.* Uppsala: Almqvist & Wiksell International.

Lorenzoni, G. and Ornati, O. A. (1988) 'Constellations of firms and new ventures', *Journal of Business Venturing* 3: 4–57.

Lundgren, A. and Björklund, L. (forthcoming) 'Economic changes in industrial networks', in B. Axelsson and G. Easton (eds) *Industrial Networks. The New Reality*, London: Routledge.

McDonald, S. (1987) 'British science parks: Reflections on the politics of high technology', *R&D Management* 17 (1): 25–37.

Malerba, F. (1985) *The Semiconductor Business. The Economics of Rapid Growth and Decline*, Madison, Wis.: University of Wisconsin Press.

Mattsson, L-G. (1969) *Integration and Efficiency in Marketing Systems*, Stockholm: EFI Stockholm School of Economics.

—— (1985) 'An application of a network approach to marketing. Defining and changing market positions', in J. Arndt and J. Dholakia (eds) *Alternative Paradigms for Widening Marketing Theory*, Greenwich, Conn.: JAI Press.

Melin, L. (1983) 'Strukturförändringar, företagsstrategier och lokala aktörer' (Changes in organization structure, company strategies and local actors), *Research Report 131*, Department of Business, University of Linköping.

Nyström, H. (1974) 'Uncertainty, information and organizational decisionmaking: a cognitive approach', *Swedish Journal of Economics* 76: 131–9.

Okimoto, D. I., Sugano, T., and Weinstein, F. B. (1984) *Competitive Edge. The Semiconductor Industry in the U.S. and Japan*, Stanford: Stanford University Press.

Patrick, H. and Meissner, L. (eds) (1986) *Japan's High Technology Industries. Lessons and Limitations of Industrial Policy*, Seattle: University of Washington Press.

Peters, L. S. (1987) *Technical Network Between U.S. and Japanese Industry*, Center for Science and Technology Policy, Rensselaer Polytechnic Institute, Troy, New York.

Pfeffer, J. and Salancik, G. R. (1978) *The External Control of Organizations. A Resource Dependence Perspective*, New York: Harper & Row.

Porter, M. E. (1985) *Competitive Advantage: Creating and Sustaining Superior Performance*, New York: The Free Press.

Powell, W. W. (1987) 'Hybrid organizational arrangements: New form or transitional development?' *California Management Review* 30 (1): 67–87.

Richardson, G. B. (1972) 'The organisation of industry', *The Economic Journal* (September): 883–96.
Rogers, E. M. (1962) *Diffusion of Innovations*, New York: The Free Press.
Rogers, E. M. and Kincaid, D. L. (1981) *Communication Networks. Towards a New Paradigm for Research*, New York: The Free Press.
Rogers, E. M. and Larsen, J. K. (1984) *Silicon Valley Fever*, New York: Basic Book.
Sahal, D. (1980) 'Technological progress and policy', in D. Sahal (ed.) *Research, Development, and Technological Innovation*, Lexington, Mass.: Lexington Books.
Sampson, A. (1976) *The Seven Sisters. The Great Oil Companies and the World They Made*, Coronet Books.
Scott, W. R. (1982) *Organizations. Rational, Natural, and Open Systems*, Englewood Cliffs, N.J.: Prentice Hall.
Sigurdson, J. (1986) 'Industry and state partnership in Japan. The very large-scale integrated circuits (VLSI) project', *Research Policy Studies*, Discussion Paper no. 168, Research Policy Institute, Lund.
Smith, P. and Easton, G. (1986) 'Network relationships: A longitudinal study', paper presented at the Third International I.M.P. Research Seminar on International Marketing, Lyon, 3–5 September 1986.
Steindl, J. (1980) 'Technical progress and evolution', in D. Sahal (ed.) *Research, Development, and Technological Innovation*, Lexington, Mass.: Lexington Books.
Takeuchi, H. and Nonaka, I. (1986) 'The new new product development game', *Harvard Business Review* (January/February): 137–46.
Thompson, J. (1967) *Organizations in Action*, Chicago: McGraw Hill.
Tung, R. (ed.) (1986) *Strategic Management in the United States and Japan. A Comparative Analysis*. Cambridge, Mass.: Ballinger Publishing Company.
Turnbull, P. W. and Valla, J-P. (eds) (1985) *Strategies for International Industrial Marketing*, London: Croom Helm.
Utterback, J. M. and Abernathy, W. J. (1975) 'A dynamic model of process and product innovation', *Omega* 3 (6): 639–56.
Utterback, J. M. and Reitberger, G. (1982) *Technology and Industrial Innovation in Sweden. A Study of New Technology-Based Firms*, Stockholm: STU.
Van de Ven, A. H., Emmit, D. G. and Koeing, R. (1975) 'Frameworks for interorganizational analysis', in A. R. Negandhi (ed.) *Interorganizational Theory*, Kent, Ohio: Kent State University Press.
von Hippel, E. (1978) 'Successful industrial products from customer ideas', *Journal of Marketing* 42 (1): 39–49.
Waluszewski, A. (1988) 'CTMP-fallet. Processutveckling inom skogsindustrin' (The CTMP case. Process development within the forest industry), Working Paper, Department of Business, University of Uppsala.
Weick, K. E. (1969) *The Social Psychology of Organizing*, Reading, Mass.: Addison Wesley.
―――― (1976) 'Educational organizations as loosely coupled systems',

Bibliography

Administrative Science Quarterly 21 (1): 1–19.
White, L., Jr. (1970) 'Technology assessment from the stance of a medieval historian', in idem., *Medieval Religion and Technology*, Berkeley, Calif.: 1978.
Williamson, O. E. (1975) *Markets and Hierarchies: Analysis and Antitrust Implications*, New York: The Free Press.
—— (1979) 'Transaction-cost economics: The governance of contractual relations', *The Journal of Law and Economics* 22 (2): 232–62.
Yoshino, M. Y. and Lifson, T. B. (1986) *The Invisible Link. Japan's Sogo Shosha and the Organization of Trade*, Cambridge, Mass.: MIT Press.

Index

Abernathy, W. J. 42
adaptations 22–3, 31–2, 66–72, 142–3
adjustments 6, 26, 32, 52, 55, 81, 104
Alchian, A. A. 27
Alderson, W. 27
Allen, T. J. 98
alliances 150
atomistic unit 34
Axelrod, R 27
Axelsson, B. 10, 27, 76, 145, 161

bank 82–4
biotechnology 38, 153–9
Björklund, L. 10
Blau, P. M. 24
Blumenthal, D. 14
bonds, types of 24

capability 16, 29, 88, 131–2
capital intensity 94–5
cellulose 161–3
chains 19, 52, 75, 99
changes, successive vs leap-wise 32, 35, 39–41, 51, 104, 120–30, 144
collaboration: costs 36, 60, 121; form 10, 37, 45; partners 35–7, 41, 56–8, 61, 70–1, 97–8, 108–12, 121, 125–6, 147; projects 26, 29, 36, 44, 60
collective action 33
combining activities and resources 31–3
commitment 24–5, 31, 120, 128

competence 89–90, 159
competitors 65, 72–5, 128–30
components 80–1
concentration, degree of 78–9
conflicts 22, 33, 128–9
consultants 44, 48, 92–3
contact structure 108, 115–16
control 18, 25–6, 29–33, 36, 134
Cook, K. S. 50
co-operation: and control 36; formal 28, 112–4; profile 37, 46, 55, 106–7, 122, 134, 147–9; propensity for 53–5
co-ordination 6, 7, 20, 22, 36, 95, 116, 149–50
corporate boundaries 10
corporate identity 51–3, 122, 132–6, 147
corporate management 131–51
creativity 32
customer-led production 54, 102–4
customers 3–4, 37, 41, 44, 48, 56–8, 65–75, 98, 109–111, 123–6, 134
Cyert, R. M. 5, 27
Czepiel, J. A. 8

defence projects 9
Demsetz, H. 27
dependency 4–5, 20, 52, 82
developed countries 7
development: activities 100–2; budget 36; internal 3–4; process 34–41; product 101; role 76; strength 25; technological 21, 29–43, 53, 120–72

Index

Dibner, M. D. 14
differentiation 70
distribution channels 27, 104–6
dynamic elements 5, 21, 23, 26, 120, 170–1

Easton, G. 14, 27
Eccles, R. G. 27
education 87–8
efficiency 23, 25, 37, 60, 67, 76–7, 149
electronics 8, 139, 161
Emerson, R. M. 50
engineering 38, 166–70
environment, geographical 38, 125
equipment 80–1, 94–5
Evan, W. M. 50
externalization 30

financial capital 17, 33, 48, 62, 82–4
Ford, D 27, 51
formalization 109, 112–4

Gadde, L-E. 27
Galeano, E. 42
geographical areas 104, 166–70
geographical location 108–11
geographically-scattered buying 58
Granstrand, O. 8
growth 59–64

Hägg, I. 27
Håkansson, H. 10, 22, 27, 30, 36, 42, 47, 50, 66, 76, 81, 145
Hamfelt, C. 98
Hammarkvist, K-O. 24, 128
Henriksson, H. 47
Hermansson, C. H. 82
heterogeneity 18, 27, 171
heterogenizing 31–2, 40
hierarchic control 18
hierarchic organization 135
hierarchization 30–2
high-technology 7–9
horizontal units 92–8, 109–11, 123–6, 134
Hörnell, E. 98
hydraulics 137–41

image processing 10, 38
industrial policy 7–9, 152–72
information processing 161
innovativeness 32
input goods 17, 48, 75–82
interactive effect 36, 41
interdependency 5, 26, 99
investments 31, 33, 59, 114, 120, 123–4, 126, 158

Jacobs, D. 27
Japan 8, 16
Johanson, J. 27, 51
joint ventures 25
just-in-time systems 24–5

Kagono, T. 27
Kincaid, D. L. 8
Kinch, N. 27
knowledge: experience 35, 38, 88–90; theory 35, 38, 88–90; transfer 144–6, 160–6

Laage-Hellman, J. 10, 50, 145, 153
language 146
Larsen, J. K. 9
laser processing 161
learning 19, 60
legal relationships 25, 37, 112–14, 126
Lifson, T. B. 14, 151
Liljegren, G. 27, 150
Lindberg, A-K. 98
Lindblom, C. E. 27, 42
Lindqvist, S. 42
Lorenzoni, G. 27
Lundgren, A. 10

McDonald, S. 8
Malerba, F. 14
March, J. G. 5, 27
market leader 58, 72–5
market shares 56–8, 72, 135
marketing resources 17, 65 ff, 131
Mattsson, L-G. 27
Meissner, L. 14
Melin, L. 27
mobilization 26, 36, 60–4, 144

Index

monopolization 33
mutuality 4, 22, 146
multiplier effect 7

national programme 8–9, 16, 152–52
network: behaviour 51–64; definition 50, 132–3; geographically-defined 45; marketing 65–75; model 16–26, 127–30; monitoring 77, 143–7; position 37, 72–3, 121, 133; processes 29–33, 132–3; structure 30–3, 75, 95, 134
Nonaka, K. 8, 27
Nyström, H. 42

oil industry 17, 27
Okimoto, D. I. 14
organizational sets 45, 50
organizing 104–106, 134, 149–50
Ornati, O. A. 27
owners 82–4

paper-pulp 10
Patrick, H. 14
personnel 17, 48, 84–92, 159
Peters, L-S. 14
Pfeffer, J. 27
political process 15, 30, 152
Porter, M. E. 27
Powell, W. W. 27
power 21, 29, 53
product range 40
production process 102–4
productivity 25
profit 46, 59–63, 148
purchasing 10, 48, 75–82, 131

questionnaire 48

rationality, local 5
rationalizations 31–2, 34, 76
raw materials 80–1
reciprocity 22
recruitment 11, 90, 150
regional difference 7 (see geographical)
Reitberger, G. 26

relationship: costs 23, 116, 121, 125, 149; customer 67–70, 98–119; developmental (definition) 11, 120; duration 22–3, 111–12, 123; features 22–6, 37; functions 25–6, 129–30; formalization 25, 37, 112–14, 126; supplier 75–82, 98–119
research and development (R&D) 153–9; companies 155, 157, 159
research institutes 3, 4, 8, 44, 48, 92, 152–72
resources: base 29, 37–8, 121, 129–30, 142–3, 147, 159; control 32, 65, 135; dependence 21, 27; development 53–8; dimensions 31–2, 45; general 16–19, 65; heterogeneous 18; human 18, 84–91; linkages 53
Richardson, G. B. 27
Rogers, E. M. 8, 9

Sahal, D. 42
Salanik, G. R. 27
Sampson, A. 27
saw mill 38
scale effect 31–2, 40, 104
science parks 8
Scott, W. R. 27
search process 5
Sigurdson, J. 14
Silicon Valley 8
Smith, P. 14, 27
Snehota, I. 66
social exchange 22–4
societal level 7, 152–72
space projects 9
specialization 36, 53, 60, 104, 109
standardization 32, 55, 70, 104
steel industry 10, 142–3
Steindl, J. 42
structuring 31–2, 40, 76, 135
sub-contractor 23, 36, 57
sub-optimization 5
supplier 3–4, 37, 41, 44, 48, 56–8, 75–82, 98, 109–11, 123–6, 134

Takeuchi, H. 8, 27

Index

technological development 21, 29–43, 53, 131–51, 152–72
technological interaction 9–10, 22–3
technology: centres 9, 166–70; dissemination 9, 160–6; fields 5, 38–9; monitoring 11, 134
Thompson, J. 27
time 60, 88–9
third party 22, 74, 128
training, technological 169
transformation 30, 32
triangle drama 74, 125, 128–30
trust 24, 115, 126, 146
Tung, R. 14
Turnbull, P. 22, 81, 115

uncertainty (genuine) 30

university 8, 38, 42, 90, 92–6, 153–72
Utterback, J. M. 26, 42

Valla, J-P. 22, 81, 115
value added 103–4
Van de Ven, A. H. 27
von Hippel, E. 8

Waluszewski, A. 10, 21, 50
Weick, K. 27
White, L, Jr. 42
Williamson, O. E. 27
wood materials 161, 163

Yoshino, M. Y. 16, 151